VOLUME EIGHTY EIGHT

ADVANCES IN PROTEIN CHEMISTRY AND STRUCTURAL BIOLOGY

VOLUME EIGHTY EIGHT

ADVANCES IN PROTEIN CHEMISTRY AND STRUCTURAL BIOLOGY

Edited by

ROSSEN DONEV

Institute of Life Sciences, College of Medicine
Swansea University, United Kingdom
Clinic and Polyclinic for Psychiatry and Psychotherapy
Rostock University, Germany

ELSEVIER

AMSTERDAM • BOSTON • HEIDELBERG • LONDON
NEW YORK • OXFORD • PARIS • SAN DIEGO
SAN FRANCISCO • SINGAPORE • SYDNEY • TOKYO
Academic Press is an imprint of Elsevier

Academic Press is an imprint of Elsevier
The Boulevard, Langford Lane, Kidlington, Oxford, OX51GB, UK
32, Jamestown Road, London NW1 7BY, UK
Radarweg 29, PO Box 211, 1000 AE Amsterdam, The Netherlands
225 Wyman Street, Waltham, MA 02451, USA
525 B Street, Suite 1900, San Diego, CA 92101-4495, USA

First edition 2012

ISBN: 978-0-12-398314-5
ISSN: 1876-1623

For information on all Academic Press publications
visit our website at store.elsevier.com

Printed and bound in USA

12 13 14 11 10 9 8 7 6 5 4 3 2 1

CONTENTS

PREFACE

Inflammation is a complex biological response to harmful stimuli such as cellular damage and pathogens. In general, it is a protective effort of the organism to remove the harmful stimuli. However, when it occurs chronically, inflammation can lead to a number of adverse effects. Inflammatory responses in the brain have been recognized for years as critical in neurodegeneration and behavior in some neurological disorders (e.g., multiple sclerosis, Alzheimer's disease). However, in recent years, researchers discovered that certain inflammatory responses are involved in correct development of the brain and in neurogenesis as well as in the protection of our brain from neurodegeneration. The presence or absence of some of the inflammatory responses in the brain often can be a result of a particular combination of polymorphisms in groups of genes controlling signaling pathways and/or protein activation cascades such as complement system. These findings suggest the important role of inflammation in neurodevelopmental disorders which has been largely overlooked so far. Now it becomes more obvious that inflammation plays a dual role in our brain and therefore further in-depth studies on the role of inflammation in neuropsychiatric disorders are urgently required. These will allow the correct design of drugs targeting the adverse effect from inflammation while maintaining the benefits for the brain.

The goal of this thematic volume of the *Advances in Protein Chemistry and Structural Biology* is to provide a comprehensive view of the mechanisms triggering inflammatory responses in the brain and leading to different neuropsychiatric disorders as well as some novel approaches for their treatment. This thematic volume provides a rationale for future studies on the relationship between inflammation and neuropsychiatric disorders with intent to inspiring the development of new agents for a more efficient management and prevention.

Rossen Donev
Institute of Life Sciences, College of Medicine, Swansea University
United Kingdom
Clinic and Polyclinic for Psychiatry and Psychotherapy
Rostock University, Germany

CHAPTER ONE

Inflammation in Anxiety

Samina Salim[1], Gaurav Chugh, Mohammad Asghar
Department of Pharmacological and Pharmaceutical Sciences, University of Houston, Houston, Texas, USA
[1]Corresponding author: e-mail address: ssalim@uh.edu

Contents

Abstract

The idea of the existence of an interaction between the immune system and the central nervous system (CNS) has prompted extensive research interest into the subject of "Psychoneuroimmunology" taking the field to an interesting level where new hypotheses are being increasingly tested. Specifically, exactly how the cross talk of pathways and mechanisms enable immune system to influence our brain and behavior is a question of immense significance. Of particular relevance to this topic is the role of cytokines in regulating functions within the CNS that ultimately modulate behavior. Interestingly, psychological stress is reported to modulate cytokine production, suggesting potential relevance of this mediator to mental health. In fact, cytokine signaling in the brain is known to regulate important brain functions including neurotransmitter metabolism, neuroendocrine function, synaptic plasticity, as well as the neural circuitry of mood. It is rather obvious to expect an aberrant behavioral outcome as a result of a dysregulation in cytokine signaling which might lead to occurrence of depression, anxiety, and cognitive dysfunction. Thus, understanding the mechanisms by which the immune system influences behavior would reveal targets for potential therapeutic development as well as strategies for the prevention of neuropsychiatric diseases. To date, the presence of inflammatory responses and the crucial role of cytokines in depression have received most attention. However, considering a big socioeconomic impact due to an alarming increase in anxiety disorder patients, there is an urgent research need for a better understanding of the role of cytokines in anxiety. In this review, we discuss recent research on the role of neuroimmunology in anxiety. At the end, we offer an "oxidative stress theory," which we propose works perhaps as a "sensor of distress," the imbalance of which leads to neuroinflammation and causes anxiety disorders. *Much research is needed to extensively test this theory keeping an open mind!*

Advances in Protein Chemistry and Structural Biology, Volume 88
ISSN 1876-1623
http://dx.doi.org/10.1016/B978-0-12-398314-5.00001-5

ABBREVIATIONS

ARB angiotensin AT1 receptor blocker
BBB blood–brain barrier
BDNF brain-derived neurotrophic factor
BSO buthionine sulfoximine
CaMK Ca^{2+}/calmodulin protein kinase
CNS central nervous system
COX cyclooxygenase
CREB cAMP response element-binding
CRP C-reactive protein
GABA gamma amino butyric acid
GLO glyoxalase
GSR glutathione reductase
HPA hypothalamic–pituitary–adrenal
IFN interferon
IL interleukin
LC locus coeruleus
MG methylglyoxal
NFκB nuclear factor kappaB
RNS reactive nitrogen species
ROS reactive oxygen species
Th1 type 1 helper
Th2 type 2 helper
TNF tumor necrosis factor
X+XO xanthine+xanthine oxidase

1. INTRODUCTION

Psychological stress is a key determinant of health and disease and reported to lead to a variety of diseases whose onset and course are now considered to be influenced by the immune system. Identified by the Egyptians, pursued by the Romans, and centuries later defined by modern researchers, as a natural immune response against stress, inflammation has become a topic of extensive research. Inflammation is considered as a key component of host defense response to injury, which is associated with adherence, and invasion of leukocytes into injured or infected tissues, and is a vital part of the immune system. Inflammation is generally considered as protective when its mechanism of action is to contain injury or infection locally. However, inflammation becomes harmful when it gets excessive (overexpression or overactivity of mediators) over time. Therefore, inflammation may have beneficial as well as detrimental actions, particularly during repair and recovery processes. For example, chronic inflammation becomes pathological upon continued active inflammatory response and extended tissue destruction. Many of the immune cells including macrophages, neutrophils, and eosinophils are involved directly or by producing inflammatory mediators in the pathology of chronic inflammation. Immune system comprises a sophisticated set of mechanisms coordinated by the interaction of a variety of specific mediators called cytokines, prostaglandins, chemokines, and others that generate nonspecific physiological responses including sickness behavior and hypothalamic–pituitary–adrenal (HPA) axis activation (Allan & Rothwell, 2003). Just as the nervous and endocrine systems convey information to the immune system via neurotransmitters and hormones, the immune system transmits information to the nervous and endocrine systems via specific mediators called cytokines and chemokines (Leonard & Myint, 2009). Cytokines are soluble mediators released both at the periphery (e.g., monocytes and macrophages) and in the brain (e.g., microglia, astrocytes, oligodendroglia, and neurons) and are associated with inflammation, immune activation, cell differentiation, and cell death (Allan & Rothwell, 2003). Cytokine production is controlled by type 1 helper (Th1) and type 2 helper (Th2) cells. Th1 cells mediate a proinflammatory cellular immune response, while Th2 cells enhance humoral immune reactions. Proinflammatory cytokines, including interleukin (IL)-1, IL-6, interferon (IFN)-γ, and tumor necrosis factor (TNF)-α, enhance the immune response to eliminate pathogens, while anti-inflammatory cytokines, including IL-4, IL-10, and IL-13, dampen the immune response by reducing the synthesis of proinflammatory cytokines (Kronfol & Remick, 2000). The balance between Th1 and Th2 therefore is important in limiting the inflammatory response (Dantzer et al., 2008), and a delicate balance of proinflammatory and anti-inflammatory cytokines is critical (Loftis, Huckans, & Morasco, 2010). In addition to exogenous immune influences, brain's own endogenous responses also impact the healthy and diseased central nervous system (CNS) and primarily comprise myeloid cells known as microglia. Microglia are macrophages that constitute the first line of innate immune defense in the brain. Usually deactivated under normal physiological conditions, microglia switch to an activated form in response to tissue damage or invasion of a pathogen and promote an inflammatory response by releasing factors that engender responses analogous to the responses of activated immune cells in the periphery (Glass, Saijo, Winner, Marchetto, & Gage, 2010). These endogenous and exogenous immune responses in the CNS do not function in isolation, rather display a dynamic interplay among themselves (Rezai-Zadeh, Gate, & Town, 2009).

For a long time, inflammation in the CNS was largely considered as bleed-over of peripheral immune responses to pathogens (viruses, bacteria, parasites) invading the CNS or an element of some type of CNS autoimmune diseases. It was generally believed that the blood–brain barrier (BBB) prevented access of immune cells to the brain and, as a result, the immune system and the CNS were believed to be relatively independent of each other. This view of inflammation is now drastically changed. It has become quite clear that BBB permeability is modulated, and trafficking of peripheral macrophages and leukocytes into the brain parenchyma occurs in a tightly regulated manner and helps promote brain homeostasis and prevent neuronal death (Ransohoff & Perry, 2009; Rezai-Zadeh et al., 2009). Thus, the CNS despite the selectivity imposed by the BBB actually responds to peripheral inflammatory stimuli and elicits a local inflammatory response called neuroinflammation. Specific routes for the movement of peripheral cytokine signals to the brain are now known (Capuron & Miller, 2011). Moreover, it also is now well established that cytokines modulate neuronal activity in specific brain regions such as the amygdala, hippocampus, hypothalamus, and the cerebral cortex (Besedovsky & del Rey, 1996; Elenkov, Wilder, Chrousos, & Vizi, 2000). It is interesting to note that these brain regions have been previously implicated in regulation of stress response (Davis, 2002). It is quite apparent that psychological stress, infection, or inflammation within the brain or the periphery can modulate cytokine expression within the CNS (Lucas, Rothwell, & Gibson, 2006). Behavioral consequences of these effects include the occurrence of several neuropsychiatric disorders (Capuron & Miller, 2011). Consequently, inflammation is now a well-recognized contributor to acute and chronic CNS disorders. In fact, neuroimmune dysregulation is believed to be responsible for the chronic elements of neurodegenerative diseases. Although it cannot be concluded with utmost certainty that neuroinflammation plays a causal role in neurodegeneration, epidemiological and preclinical data suggest that chronic neuroinflammation triggers neuronal dysfunction during the asymptomatic stage of neurodegenerative diseases including Parkinson's and Alzheimer's diseases. Moreover, evidence from medical studies implicates the immune system in a number of psychiatric disorders with developmental origins, including schizophrenia, anxiety, depression, and cognitive dysfunction. Several excellent reviews have discussed in depth many important aspects of CNS inflammation, with regard to neurodegeneration and neuropsychiatric ailments (Gebicke-Haerter, 2001; Nguyen, Julien, & Rivest, 2002; Perry, Bolton, Anthony, & Betmouni, 1998). Clearly, the most studied and documented area has been major depression. Ever since the involvement of immune function in depression was first reported (Maes et al., 1990, 1991, 1992a, 1992b), numerous studies examining the role of cytokines in major depression have been reported. However, research data and literature reviews on anxiety and neuroinflammation are particularly lacking. Hou and Baldwin (2012) and Hovatta, Juhila, and Donner (2011) recently presented an excellent review of anxiety–inflammation literature. The purpose of this review is not to revisit the same ground, but to present an overview of the critical features of inflammation and to reveal mechanisms potentially critical to anxiety disorders in particular. Limitations of current approaches and gaps in our knowledge as well as implications of recent discoveries also are discussed. Finally, we will speculate on the role and therapeutic potential of some targets of the inflammatory cascade that may be critical to the understanding of anxiety disorders.

2. CYTOKINES, BRAIN, AND BEHAVIOR

Cytokines are proteins, peptides, or glycoproteins released by cells that serve as cellular signals to regulate immune response to injury and infection. Initially described as immune cell mediators in the periphery, they are also expressed in the CNS and are involved in the modulation of various neurological functions (Viviani, Bartesaghi, Corsini, Galli, & Marinovich, 2004; Viviani & Boraso, 2011; Viviani, Gardoni, & Marinovich, 2007). In the brain, cytokines work as an integrated network by inducing their own synthesis as well as that of other proinflammatory cytokines. Cytokine signals originating in periphery reach the brain through humoral, neural, and cellular pathways (Capuron & Miller, 2011). Cytokines are classified as proinflammatory and anti-inflammatory. The proinflammatory cytokines include IL-1, IL-6, and TNF which promote inflammation, a beneficial effect in early immune responses to infection or injury (Glaser, Rabin, Chesney, Cohen, & Natelson, 1999). The primary purpose of cytokines is to attract immune cells to the site of infection or injury and activating them to respond. Other actions secondary to these include changes in physiology to promote inflammation, like alterations in metabolism and temperature regulation. Anti-inflammatory cytokines such as IL-10 and IL-13 dampen the immune response, causing, for instance, restoration of cellular function and inhibition of proinflammatory cytokine synthesis. The immune system's inflammatory response can be triggered in a variety of ways, including infection and trauma. Other commonly known inflammatory mediators include complement adhesion molecules, cyclooxygenase (COX) enzymes, and their products. Proinflammatory cytokines, including IL-1, IL-6, and TNF-α, are released not only by activated immune cells during the host response to pathogen invasion or during tissue injury but also upon psychological stress. Cytokines are known to mediate the communication between the immune system and the brain, and seem to coordinate cellular response to immune challenges, and enable behavioral changes needed for recovery. The release of proinflammatory cytokines during an immune response is generally transient and regulated by anti-inflammatory mechanisms. Thus, behavioral effects initiated by the activation of the inflammatory response develop as temporary and controlled reaction of the CNS to immune signals. However, when immune challenge becomes chronic, as that observed in patients with chronic medical illness and/or facing persistent psychological stress, the behavioral effects of cytokines and the resultant inflammatory response may contribute to the development of neuropsychiatric diseases. There is abundant information regarding the pathways and mechanisms via which the immune system potentially influences brain and behavior. Role of proinflammatory cytokines in regulating functions in the CNS is well known. Interestingly, psychological stress has been shown to increase cytokine production. Cytokine signaling in the brain is known to regulate important brain functions including neurotransmitter metabolism, neuroendocrine function, synaptic plasticity, and mood-regulating neural circuitry. The behavioral outcome of any type of dysregulation of the immune system in the CNS might lead to occurrence of depression, anxiety, cognitive dysfunction, and sleep impairment.

Considering the tremendous attention generated with regard to inflammation as a causal factor for many illnesses including cardiovascular disease, diabetes, and cancer, as well as its role in depression and anxiety disorders, it is imperative to identify specific molecular targets of immune function that may be critical for novel immune-based therapeutics for mental health disorders. Tremendous amount of work has been done with regard to depression. Involvement of cytokines and inflammatory factors in the pathophysiology of depression has been proposed by several groups (Dantzer, O'Connor, Freund, Johnson, & Kelley, 2008; Miller, Maletic, & Raison,

2009; Raison, Capuron, & Miller, 2006). Increased prevalence of comorbidity between depression and chronic inflammatory diseases like rheumatoid arthritis, cancer, infectious diseases, autoimmune, and cardiovascular disease (Evans et al., 2005) lends further support to this proposition. In depressed patients, TNF-α, IL-6, and C-reactive protein (CRP) are reported to be elevated in several studies (Dowlati et al., 2010; Howren, Lamkin, & Suls, 2009; Zorrilla et al., 2001), and a convincing role of inflammation in the pathophysiology of depression has been reported by several groups (Capuron & Miller, 2004; Musselman et al., 2001; Raison et al., 2005). In particular, elevated inflammatory markers including proinflammatory cytokines, chemokines, and adhesion molecules have been reported in the blood and cerebrospinal fluid of patients with major depression (Dowlati et al., 2010; Howren et al., 2009; Miller et al., 2009; Zorrilla et al., 2001). Significant correlation has also been described between markers of inflammation and neuropsychiatric symptoms including fatigue and cognitive dysfunction (Bower et al., 2009; Jehn et al., 2006; Lutgendorf et al., 2008; Meyers, Albitar, & Estey, 2005; Musselman et al., 2001). Furthermore, alterations in immune function have been found in patients with major depression. Reports of immune suppression (e.g., reduced natural killer cell activity and reduced lymphocyte proliferation) followed by increased inflammatory activity (e.g., increased circulating levels of inflammatory markers) have been described (Anisman, Ravindran, Griffiths, & Merali, 1999; Irwin & Gillin, 1987; Kronfol & Remick, 1983; Maes et al., 1993; Zorrilla et al., 2001). During an infection, the proinflammatory response that is essential for an active immune defense is normally contained by cortisol and also by parasympathetic activity (Borovikova et al., 2000; Munck & Guyre, 1991). Yet, inadequate containment often leads to septic shock and death, and treatment with cortisol and elevation of parasympathetic activity are two pathways to reduce an excessive inflammatory response (Munck & Guyre, 1991). At the opposite extreme, too much cortisol can compromise immune defenses by suppressing the proinflammatory responses (Munck & Guyre, 1991; Sapolsky, Romero, & Munck, 2000). These two examples of too much or too little activity of certain mediators of allostasis illustrate allostatic overload (McEwen & Wingfield, 2003). Allostatic overload is generally defined as the wear and tear produced by imbalances in allostatic mediators. Some other examples of allostatic overload include conditions like hypertension, atherosclerosis, diabetes, and metabolic syndrome, and stress-induced remodeling in brain regions which determine learning and memory functions and anxiety (McEwen, 2004; McEwen & Wingfield, 2003). Such changes in brain structure are seen in major depression and Cushing's disease.

3. INFLAMMATION IN ANXIETY DISORDERS

Although depression has received most attention, relevance of inflammation to other neuropsychiatric diseases including anxiety disorders is inevitable considering the impact that cytokines have on neurotransmitter systems and associated circuitry related to anxiety response. However, due to the increased comorbidity between depression and anxiety, it becomes incredibly difficult to separate cause–effect relationship between the two especially when considering the involvement of a third component such as neuroinflammation. It is likely that critical brain–immune system interactions occur that are unique to each condition and regulate the biochemistry within specific brain areas that determine depression and anxiety differently. Some suggest that, since antidepressants have a protective effect on both anxiety and depression, similar neurobiological substrates are affected. A complete overlap of biochemistry within the CNS in the two disorders seems unlikely. While psychological chronic stress, by causing changes in the HPA and the immune system, can elicit anxious and depressive behaviors simultaneously (Leonard & Myint, 2009), the possibility that different circuits are activated and novel signaling cascades are involved is an intriguing possibility. This scenario makes lot of sense when one considers the data suggesting higher levels of IL-6 in people with anxiety, independent of depressive symptoms suggesting an anxiety-specific effect on inflammatory response. Perhaps this is one of the pathways via which anxiety increases the risk for other inflammatory conditions (O'Donovan et al., 2010). Whether anxiety is causal for inflammation or inflammation elicits anxiety response is an intriguing question. Our position on this issue is that perhaps inflammation mediates critical biochemical changes within specific areas of the brain which leads to anxiety phenotype. Some of this biochemistry will be addressed below.

Research in the past several years has shown quite interesting results. Several animal studies suggest a potentially important role of neuroinflammation in anxiety. For example, increased cytokine expression in the periphery was reported to be associated with heightened anxiety-like behavior in mice (Sakic et al., 1994; Schrott & Crnic, 1996). Moreover, mice overexpressing IL-6 or TNF developed an anxiogenic phenotype (Connor & Leonard, 1998; Fiore et al., 1998). Interestingly, deletion of the gene encoding IFN-γ enhances anxiety-like behavior in rodents (Fiore et al., 1998; Kustova, Sei, Morse, & Basile, 1998; Lesch, 2001). As far as the human studies are concerned, there is considerable amount of data showing that high level of anxiety is associated with impaired cellular immunity (Boscarino, 2004; Godbout & Glaser, 2006; Schneiderman, Ironson, & Siegel, 2005; Zhou et al., 2005), including damage to cellular and humoral immune responses (Arranz, Guayerbas, & De la Fuente, 2007; Koh & Lee, 2004; Zhou et al., 2005) and increased vulnerability to infections (Aviles, Johnson, & Monroy, 2004; Takkouche, Regueira, & Gestal-Otero, 2001). Normal volunteers showed anxiety symptoms when injected with the immune activator lipopolysaccharide (Reichenberg et al., 2001). Several groups have reported a positive correlation between anxiety and inflammatory markers in humans (TNFα, IL-6, and CRP) (Arranz et al., 2007; Maes et al., 1998; Pitsavos et al., 2006). However, studies of Zorrilla, Redei, and DeRubeis (1994) have demonstrated a negative correlation. The inconsistency calls for a more extensive investigation including a larger sample size and a uniform sample selection excluding gender differences and other preexisting conditions.

Moving on the theme of involvement of unique neuroinflammatory cell-signaling cascades potentially critical to anxiety disorders, below we will discuss the role of oxidative stress within the CNS in regulation of anxiety phenotype.

4. INFLAMMATION IN ANXIETY DISORDERS: POTENTIAL ROLE OF OXIDATIVE STRESS

Oxidative stress is a state where the level of oxidants (hydrogen peroxide, superoxide, nitric oxide, etc.) produced by biological reactions exceeds the oxidants scavenging capacity of the cells. These oxidants modify cellular macromolecules (proteins, DNA, lipids) and alter cellular functions (Raha & Robinson, 2001). Increased oxidative damage most probably occurs in most if not all human diseases, although it perhaps plays a causal role only in a few. The role of oxidative stress in the development of neurodegenerative disorders is well reported (Greco & Fiskum, 2010; Lassmann, Brück, & Lucchinetti, 2001). Several theories have been proposed over the years to conceptualize the pathophysiology of anxiety disorders. Those theories include mechanisms such as impairment of neurotransmission, genetic mutations, excitotoxicity, and stress. While most classical theories suggest involvement of traditional signal transduction mechanisms including abnormalities in the gamma amino butyric acid (GABA) and serotonin receptor systems in the etiology of anxiety, recent work from several labs (Bouayed, Rammal, & Soulimani, 2009; Bouayed, Rammal, Younos, & Soulimani, 2007; de Oliveira, Silvestrin, Mello, Souza, & Moreira, 2007; Hovatta et al., 2005; Masood, Nadeem, Mustafa, & O'Donnell, 2008; Souza et al., 2007) including our own published findings, has produced some provocative and interesting results, suggesting relevance of oxidative stress to anxiety disorders and implying that perhaps oxidative stress is the new stress as has been suggested previously (Gingrich, 2005).

Oxidative stress has been suggested to act as an initiator and/or a mediator of several human diseases. Increased oxidative and nitrosative stress triggered via generation of reactive oxygen species (ROS) and reactive nitrogen species (RNS), respectively, is reported to occur in several disorders of the CNS (Calabrese, Bates, & Stella, 2000; Halliwell, 2006; Sayre, Perry, & Smith, 2008). This association is largely due to the high vulnerability of brain to oxidative load (Ng, Berk, Dean, & Bush, 2008). Relevant to this, oxidative stress is reported to trigger chronic neuroinflammation, which is characterized by the generation of proinflammatory mediators locally produced by host cells, thus engaging the innate immune system. Both resident CNS cells and recruited leukocytes express various cytokines, major histocompatibility complex, and adhesion/costimulatory molecules, which collectively cause the generation of ROS (Floyd, 1999a, 1999b; Floyd et al., 1999; Hopkins & Rothwell, 1995; Martiney, Cuff, Litwak, Berman, & Brosnan, 1998; Merrill & Benveniste, 1996; Wakita, Shintani, Yagi, Asai, & Nozawa, 2001; Xiao & Link, 1998). ROS are highly toxic in the CNS when their production exceeds the neutralizing effects of endogenous antioxidants. Thus, the balance between pro- and anti-inflammatory factors determines the intensity and course of the inflammatory response including the levels of oxidative stress and consequent neurodegeneration (Bethea, 2000; Neumann, 2000; Sternberg, 1997; Stoll, Jander, & Schroeter, 2000; Streit, Walter, & Pennell, 1999). Neuroinflammation has been proposed to promote oxidative stress and contribute to irreversible neuronal dysfunction and cell death (Maccioni, Rojo, Fernandez, & Kuljis, 2009; Mhatre, Floyd, & Hensley, 2004), although neuroinflammation can be a cause as well as a consequence of chronic oxidative stress. Increased inflammation is reported to generate reactive oxygen and RNS in ambient neurons (Apelt, Bigl, Wunderlich, & Schliebs, 2004; Mhatre et al., 2004). The proinflammatory cytokines IL-1β, IL-6, and TNF-α elicit immune responses in the CNS during inflammation (Hopkins & Rothwell, 1995; Martiney et al., 1998; Xiao & Link, 1998). They are potentially deleterious or beneficial, depending on their concentration, site, and duration of action (Bethea, 2000; Hermann, Rogers, Bresnahan, & Beattie, 2001; Longhi et al., 2001; Merrill & Benveniste, 1996; Sternberg, 1997; Stoll et al., 2000). Hence, excess or prolonged production of proinflammatory cytokines, such as in genetically manipulated mice, can lead to chronic inflammation and functional CNS-related disorders (Campbell, 1998a, 1998b; Campbell et al., 1993; Carrasco et al., 2000; Giralt et al., 2001; Probert, Akassoglou, Pasparakis, Kontogeorgos, & Kollias, 1995). New data show that direct induction of oxidative stress in male Sprague–Dawley rats via two separate oxidative stress inducers, X+XO (xanthine+xanthine oxidase) and BSO (buthionine sulfoximine), leads to increased CRP-1 and IL-6 in the serum and elevated TNF-α in the brain tissues of these rats (Salim et al., 2011). Interestingly, these rats have been reported to exhibit anxiety-like behavior in open-field and light–dark anxiety-like behavior tests (Salim et al., 2010a, 2010b). Anxiety-like behavior is linked to oxidative stress status not only in cerebral system but also in peripheral system in anxious mice. Rammal, Bouayed, Younos, and Soulimani (2008) report an imbalance in oxidative status in both neuronal and glial cells in the cerebellum and hippocampus, neurons of the cerebral cortex, and peripheral leucocytes in anxious mice and suggest a potential role of this redox system in the development of neuroinflammation and neurodegeneration. Relevant to this, activation of inflammatory pathways has been observed in patients with anxiety disorders (O'Donovan et al., 2010). Increased proinflammatory cytokines have also been detected in patients with major depression and anxiety disorders (Bob et al., 2010; O'Donovan et al., 2010). Particularly, increased IL-6 in depression (Bob et al., 2010) and anxiety disorder patients (O'Donovan et al., 2010) also has been reported. Multiple sclerosis is a disease of the CNS characterized by chronic inflammation and degenerative changes (Compston & Coles, 2008). Not surprisingly, the rate of affective disorders such as depression and anxiety is at least sixfold increased in this condition. In an animal model that mimics many aspects of multiple sclerosis (myelin oligodendrocyte glycoprotein experimental autoimmune encephalomyelitis), an increased anxiety-like behavior was displayed, which correlated with increase in hippocampal tissue TNF-α levels and neuronal loss (Peruga et al., 2011).

Another line of evidence comes from chronic fatigue syndrome, which is an animal model of chronic stress characterized by persistent fatigue, infection, and rheumatological and neuropsychiatric symptoms (Chambers, Bagnall, Hempel, & Forbes, 2006) accompanied with increased oxidative stress and inflammation (Kumar, Kumari, & Kumar, 2010). Use of anti-inflammatory cyclooxygenase inhibitors (COX; both selective and nonselective) in mice with chronic fatigue (forced swimming for 15 days)-induced anxiety-like behavior and increased oxidative stress significantly attenuates these behavioral and biochemical alterations (Kumar et al., 2010). COX inhibitors demonstrate similar protective benefits in an acute stress paradigm of anxiety-like behavior and oxidative damage. Acute stress is also known to produce several behavioral, neurochemical, and biochemical alterations. Pretreatment with COX inhibitors in mice subjected to 6 h acute immobilization stress significantly reduced anxiety and oxidative stress (Kumari, Kumar, & Dhir, 2007). Similarly, Chen et al. (2011) in their study in sleep-disturbed hemodialysis patients suggest that while cognitive-behavioral therapy is effective in correcting disorganized sleep patterns, it also reduces inflammation, oxidative stress, and anxiety. Perhaps, it is likely that enhanced proinflammatory cytokine signaling may promote ROS generation and lead to oxidative damage, and this might be one mechanism that links inflammation to neuropsychiatric diseases. Excessive angiotensin II AT1 receptor activity is also associated with increased peripheral and brain inflammation, which is usually normalized by administering angiotensin AT1 receptor blockers (ARBs) (Saavedra, Sánchez-Lemus, & Benicky, 2011). ARBs are generally used for treating cardiovascular and metabolic disorders where inflammation is a major pathogenic factor (Barra, Vitagliano, Cuomo, Vitagliano, & Gaeta, 2009; Savoia & Schiffrin, 2007); however, an increasing body of evidence now suggests their extended antianxiety and antidepressant properties (Gard, Haigh, Cambursano, & Warrington, 2001; Kaiser et al., 1992; Saavedra et al., 2006; Shekhar et al., 2006). Anxiety-like behavior is also accompanied by an increase in oxidative stress in a rat model of fetal alcohol spectrum disorders. In a study by Brocardo et al. (2012), perinatal ethanol exposure (throughout gestation and during first 10 days postnatal) resulted in an anxiety-like behavior with concomitant increase in the levels of lipid peroxidation and protein oxidation in the hippocampus and cerebellum, and reduction in antioxidant glutathione levels in ethanol-exposed adult Sprague–Dawley rats. Depression is another CNS disorder that is frequently reported to be associated with immune activation and oxidative stress. This ability of depression to contribute toward increased neuroinflammatory burden has been proposed by Maes, Kubera, Obuchowiczwa, Goehler, and Brzeszcz (2011) as an explanation for the co-occurrence of anxiety with depression. In an interesting but contrasting study by Blackmore et al. (2011), the hypothesis that proinflammatory cytokines may be involved in the pathophysiology of psychiatric symptoms particularly anxiety, depression, and trauma was investigated in pregnant women. The researchers found no evidence of a significant association between anxiety and proinflammatory cytokines (IL-6 and TNF-α), suggesting that the hypothesis may not be extended to psychiatric symptoms during pregnancy.

5. OXIDATIVE STRESS, TRANSCRIPTION FACTORS, AND INFLAMMATION

While increased levels of proinflammatory cytokines seem to play an important role in neuroinflammation-mediated behavioral consequences including anxiety-like behaviors, activation of inflammation-related transcription factors, such as nuclear factor kappa B (NFκB) and cAMP response element-binding (CREB), is an interesting target to consider. NFκB is activated upon oxidative stress and reported to intensify oxidative stress leading to inflammation and neuronal damage via induction of proinflammatory cytokines (Elks et al., 2009; Mattson & Camandola, 2001). Activation of NFκB detected in selected brain areas including hippocampus, amygdala, and locus coeruleus (LC) upon induction of direct oxidative stress has also been reported in rats that display anxiety-like behavior (Salim et al., 2011). Causal role of NFκB or CREB in this event remains to be investigated. NFκB is a heterodimer of two subunits, p65 and p50 (Liou & Baltimore, 1993). NFκB p50-deficient mice have been reported to show reduced anxiety-like behaviors (Kassed & Herkenham, 2004). Another inflammation-linked transcription factor of significance to the present discussion is CREB which is widely known to promote neuronal survival, particularly through the transcriptional activation of prosurvival factors (Lonze & Ginty, 2002; Mantamadiotis et al., 2002). In fact, CREB is reported to mediate some of the anxiolytic effects of 5-HT1A serotonin receptor agonists and selective serotonin reuptake inhibitors (Boer et al., 2010). Reduced levels of Ca^{2+}/calmodulin protein kinase (CaMK)-IV and CREB in the brain have been detected in two rodent models of high inflammation, exhibiting oxidative stress-mediated anxiety-like behavior (Salim et al., 2011). The expression of these transcription factors becomes all the more intriguing to this discussion as oxidative stress is known to lead to accumulation of advanced glycation end-products (Thornalley, 2005) and contribute to cytotoxicity and inflammation. An antioxidant defense network comprising an antioxidant system including GLO-GSH is known to detoxify dicarbonylation (Thornalley, 2003) and methylglyoxal (MG)-mediated cytotoxicity (Di Loreto et al., 2008). Relevant to this, brain-derived neurotrophic factor (BDNF) is considered protective against MG challenge and believed to protect neurons from oxidative stress and to promote neurogenesis (Duman, Heninger, & Nestler, 1997). An antioxidant role of BDNF also has been proposed (Chan, Wu, Chang, Hsu, & Chan, 2010), and BDNF has been proposed to be critical to the pathophysiology of anxiety disorders (Martinowich, Manji, & Lu, 2007). Moreover, increased intraneuronal calcium in the brain is known to follow oxidative stress and also is reported to activate calcium-activated proteases, the calpains (Ghosh & Greenberg, 1995; Kim et al., 2007). CaMK-IV is a substrate for calpain (Tremper-Wells & Vallano, 2005) and catalyzes the phosphorylation of CREB (Corcoran & Means, 2001). CREB regulates BDNF gene expression (Einat et al., 2003; Pandey, 2003). This pathway involving calpains, CaMKIV and CREB, therefore seems quite relevant to regulation of BDNF. Moreover, activation of inflammatory pathways has been observed in patients with anxiety disorders (O'Donovan et al., 2010), and NFκB is activated upon oxidative stress and also known to enhance oxidative stress causing inflammation and leading to neuronal damage (Elks et al., 2009; Mattson & Camandola, 2001). Calpain-dependent NFκB activation is also known (Shumway, Maki, & Miyamoto, 1999).

Interestingly, reduced levels of antioxidant enzymes glyoxalase (GLO)-1, glutathione reductase (GSR)-1, and BDNF in hippocampus, amygdala, and LC were reported upon subchronic induction of oxidative stress with two separate oxidative stress inducers, BSO, or with X+XO treatments. Involvement of these brain areas in anxiety disorders is known (Davis, 2002; Dell'Osso, Buoli, Baldwin, & Altamura, 2010; Lopez, Akil, & Watson, 1999; Shin, Rauch, & Pitman, 2006), and their response to inflammation is also known (Besedovsky & del Rey, 1996; Elenkov et al., 2000), but their role in regulation of anxiety or inflammation is unclear (Menard & Treit, 1999; Shin et al., 2006). Furthermore, subchronic induction of oxidative stress increased the expression of calpains in the hippocampus, amygdala, and LC, accompanied with reduced expression of CaMKIV and CREB. CaMKIV is a substrate of calpain and known to regulate the expression of CREB which is, in turn, reported to modulate BDNF expression (Einat et al., 2003; Pandey, 2003). Perhaps BDNF levels are downregulated due to calpain-dependent degradation of CaMKIV and CREB, its upstream regulators. These observations are quite interesting considering that oxidative stress causes protein glycation and GLO-GSH system detoxifies dicarbonylation (Thornalley, 2003), and BDNF is upregulated upon MG challenge. Perhaps BDNF levels increase during acute oxidative stress (3-day BSO treatment) as an immediate protective response against glycation. Due to this compensatory response, antioxidant defense (GLO1 and GSR1 expression) also is upregulated. However, upon induction of chronic oxidative stress (7-day BSO or 7-day X+XO treatment), the adaptive mechanism is lost due to calpain-dependent degradation of BDNF. Therefore, it seems reasonable to postulate that reduction in antioxidant enzymes GLO1 and GSR1 leads to excessive protein glycation. BDNF is unable to compensate for this defect, and impairment in this detoxification mechanism causes drastic and irreversible alterations in the antioxidant homeostasis in the brain. Furthermore, BDNF is not only considered as protective against oxidative stress but is also known to be a prosurvival factor and believed to promote neurogenesis (Duman et al., 1997) and synaptic plasticity (Schaaf, Workel, Vreugdenhil, Oitzl, & de Kloet, 2001). Thus, it raises an interesting possibility that decreased BDNF favors a condition of high oxidative stress, high inflammation, and reduced synaptic plasticity, and all of these eventually lead to a potentially high anxiety state. Role of several transcription factors in regulation of expression of many inflammation-related enzymes including NOS, COX2, and NADPH oxidase (superoxide producing enzyme) to enhance the production of ROS is being increasingly explored. Of these, the NADPH oxidase pathway is reported to be associated with the regulation of anxiety-like behavior in mice (Masood et al., 2008). Thus, it is reasonable to expect that increases in toxic factors (inflammation, cytokines, oxidative stress) and decreases in neuroprotective factors (BDNF and vascular endothelial growth factor) yield a plausible set of rationales for the increase in major CNS structural modifications including in the hippocampus, amygdala, and LC, and this damage has pathophysiological consequences inducing anxiety disorders.

6. CONCLUSION

Research conducted over the past several years has established that important interactions occur between the immune system and the CNS and that these interactions lead to specific behavioral changes. But to move beyond this "interaction theory" and to gain critical insights into the outcome of these interactions specifically with regard to anxiety disorders, a shift toward nontraditional approaches is needed. Our working model (as shown in Fig. 1.1) is that persistent psychological stress increases oxidative stress which causes impairment of antioxidant defense by modulating specific antioxidant enzymes, leading to excessive dicarbonyl glycation. This activates calpain expression which degrades BDNF and causes impairment in detoxification processes. This contributes to even greater oxidative stress, reducing neurogenesis and plasticity, and promotes NFκB-dependent inflammation. Chronically, elevated levels of cytokines prolong central inflammatory responses by increasing BBB permeability to peripheral immune cell traffic creating an environment of even greater inflammatory and oxidative stresses accelerating oxidative damage to the CNS eliciting adverse behavioral

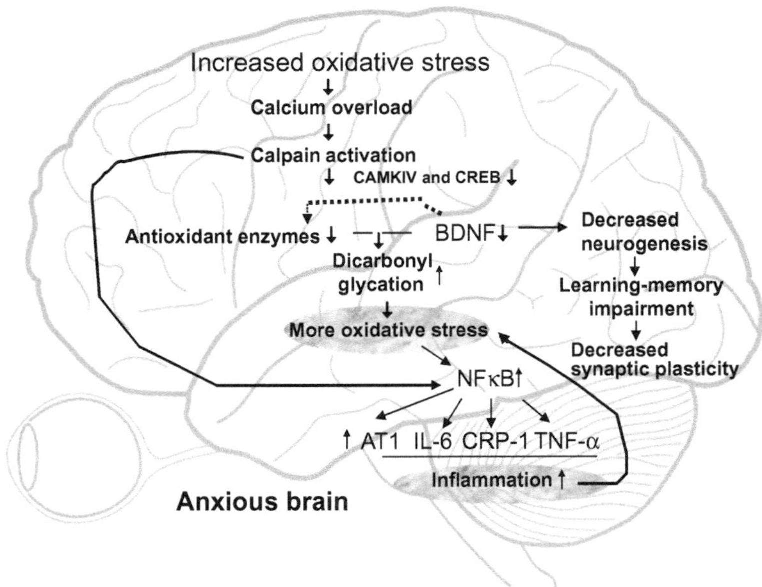

Figure 1.1 Schematic representation of putative mechanisms involved in development of anxiety.

consequences. These are attractive concepts albeit in their embryonic stages. More research is needed for a deeper understanding into the mechanisms of inflammation–oxidative stress-mediated brain damage which is potentially causal for anxiety. Targeting this area of research holds great promise for reducing/treating incidence of anxiety disorders.

ACKNOWLEDGMENTS

Grants to Enhance and Advance Research (GEAR) from University of Houston and NIH/NIMH Grant # G103327 (1R15MH093918-01A1) to S. S. and NIH/NIA Grant #AG039836 to M. A.

REFERENCES

Allan, S. M., & Rothwell, N. J. (2003). Inflammation in central nervous system injury. *Philosophical Transactions of the Royal Society of London. Series B, Biological Sciences, 358*, 1669–1677.

Anisman, H., Ravindran, A. V., Griffiths, J., & Merali, Z. (1999). Endocrine and cytokine correlates of major depression and dysthymia with typical or atypical features. *Molecular Psychiatry, 4*, 182–188.

Apelt, J., Bigl, M., Wunderlich, P., & Schliebs, R. (2004). Aging-related increase in oxidative stress correlates with developmental pattern of beta-secretase activity and beta-amyloid plaque formation in transgenic Tg2576 mice with Alzheimer-like pathology. *International Journal of Developmental Neuroscience, 22*, 475–484.

Arranz, L., Guayerbas, N., & De la Fuente, M. (2007). Impairment of several immune functions in anxious women. *Journal of Psychosomatic Research, 62*, 1–8.

Aviles, H., Johnson, M. T., & Monroy, F. P. (2004). Effects of cold stress on spleen cell proliferation and cytokine production during chronic Toxoplasma gondii infection. *Neuroimmunomodulation, 11*, 93–102.

Barra, S., Vitagliano, A., Cuomo, V., Vitagliano, G., & Gaeta, G. (2009). Vascular and metabolic effects of angiotensin II receptor blockers. *Expert Opinion on Pharmacotherapy, 10*, 173–189.

Besedovsky, H. O., & del Rey, A. (1996). Immune-neuro-endocrine interactions: Facts and hypotheses. *Endocrine Reviews, 17*, 64–102.

Bethea, J. R. (2000). Spinal cord injury-induced inflammation: A dual-edged sword. *Progress in Brain Research, 128*, 33–42.

Blackmore, E. R., Moynihan, J. A., Rubinow, D. R., Pressman, E. K., Gilchrist, M., & O'Connor, T. G. (2011). Psychiatric symptoms and proinflammatory cytokines in pregnancy. *Psychosomatic Medicine, 73*, 656–663.

Bob, P., Raboch, J., Maes, M., Susta, M., Pavlat, J., Jasova, D., et al. (2010). Depression, traumatic stress and interleukin-6. *Journal of Affective Disorders, 120*, 231–234.

Boer, U., Noll, C., Cierny, I., Krause, D., Hiemke, C., & Knepel, W. (2010). A common mechanism of action of the selective serotonin reuptake inhibitors citalopram and fluoxetine: Reversal of chronic psychosocial stress-induced increase in CRE/CREB-directed gene transcription in transgenic reporter gene mice. *European Journal of Pharmacology, 633*, 33–38.

Borovikova, L. V., Ivanova, S., Zhang, M., Yang, H., Botchkina, G. I., Watkins, L. R., et al. (2000). Vagus nerve stimulation attenuates the systemic inflammatory response to endotoxin. *Nature, 405*, 458–462.

Boscarino, J. A. (2004). Posttraumatic stress disorder and physical illness: Results from clinical and epidemiologic studies. *Annals of the New York Academy of Sciences, 1032*, 141–153.

Bouayed, J., Rammal, H., & Soulimani, R. (2009). Oxidative stress and anxiety: Relationship and cellular pathways. *Oxidative Medicine and Cellular Longevity*, *2*, 63–67.
Bouayed, J., Rammal, H., Younos, C., & Soulimani, R. (2007). Positive correlation between peripheral blood granulocyte oxidative status and level of anxiety in mice. *European Journal of Pharmacology*, *564*, 146–149.
Bower, J. E., Ganz, P. A., Tao, M. L., Hu, W., Belin, T. R., Sepah, S., et al. (2009). Inflammatory biomarkers and fatigue during radiation therapy for breast and prostate cancer. *Clinical Cancer Research*, *15*, 5534–5540.
Brocardo, P. S., Boehme, F., Patten, A., Cox, A., Gil-Mohapel, J., & Christie, B. R. (2012). Anxiety- and depression-like behaviors are accompanied by an increase in oxidative stress in a rat model of fetal alcohol spectrum disorders: Protective effects of voluntary physical exercise. *Neuropharmacology*, *62*, 1607–1618.
Calabrese, V., Bates, T. E., & Stella, A. M. (2000). NO synthase and NO-dependent signal pathways in brain aging and neurodegenerative disorders: The role of oxidant/antioxidant balance. *Neurochemical Research*, *25*, 1315–1341.
Campbell, I. L. (1998a). Structural and functional impact of the transgenic expression of cytokines in the CNS. *Annals of the New York Academy of Sciences*, *840*, 83–96.
Campbell, I. L. (1998b). Transgenic mice and cytokine actions in the brain: Bridging the gap between structural and functional neuropathology. *Brain Research. Brain Research Reviews*, *26*, 327–336.
Campbell, I. L., Abraham, C. R., Masliah, E., Kemper, P., Inglis, J. D., Oldstone, M. B., et al. (1993). Neurologic disease induced in transgenic mice by cerebral overexpression of interleukin 6. *Proceedings of the National Academy of Sciences of the United States of America*, *90*, 10061–10065.
Capuron, L., & Miller, A. H. (2004). Cytokines and psychopathology: Lessons from interferon-alpha. *Biological Psychiatry*, *56*, 819–824.
Capuron, L., & Miller, A. H. (2011). Immune system to brain signaling: Neuropsychopharmacological implications. *Pharmacology & Therapeutics*, *130*, 226–238.
Carrasco, J., Giralt, M., Penkowa, M., Stalder, A. K., Campbell, I. L., & Hidalgo, J. (2000). Metallothioneins are upregulated in symptomatic mice with astrocyte-targeted expression of tumor necrosis factor-alpha. *Experimental Neurology*, *163*, 46–54.
Chambers, D., Bagnall, A. M., Hempel, S., & Forbes, C. (2006). Interventions for the treatment, management and rehabilitation of patients with chronic fatigue syndrome/myalgic encephalomyelitis: An updated systematic review. *Journal of the Royal Society of Medicine*, *99*, 506–520.
Chan, S. H., Wu, C. W., Chang, A. Y., Hsu, K. S., & Chan, J. Y. (2010). Transcriptional upregulation of brain-derived neurotrophic factor in rostral ventrolateral medulla by angiotensin II: Significance in superoxide homeostasis and neural regulation of arterial pressure. *Circulation Research*, *107*, 1127–1139.
Chen, H. Y., Cheng, I. C., Pan, Y. J., Chiu, Y. L., Hsu, S. P., Pai, M. F., et al. (2011). Cognitive-behavioral therapy for sleep disturbance decreases inflammatory cytokines and oxidative stress in hemodialysis patients. *Kidney International*, *80*, 415–422.
Compston, A., & Coles, A. (2008). Multiple sclerosis. *Lancet*, *372*, 1502–1517.
Connor, T. J., & Leonard, B. E. (1998). Depression, stress and immunological activation: The role of cytokines in depressive disorders. *Life Sciences*, *62*, 583–606.
Corcoran, E. E., & Means, A. R. (2001). Defining Ca2+/calmodulin-dependent protein kinase cascades in transcriptional regulation. *The Journal of Biological Chemistry*, *276*, 2975–2978.
Dantzer, R., Capuron, L., Irwin, M. R., Miller, A. H., Ollat, H., Perry, V. H., et al. (2008). Identification and treatment of symptoms associated with inflammation in medically ill patients. *Psychoneuroendocrinology*, *33*, 18–29.

Dantzer, R., O'Connor, J. C., Freund, G. G., Johnson, R. W., & Kelley, K. W. (2008). From inflammation to sickness and depression: When the immune system subjugates the brain. *Nature Reviews. Neuroscience*, *9*, 46–56.
Davis, K. L. (2002). Neuropsychopharmacology: The fifth generation of progress. In (5th ed.). Philadelphia, PA: Lippincott Williams & Wilkins (LWW) Ch 63.
de Oliveira, M. R., Silvestrin, R. B., Mello, E., Souza, T., & Moreira, J. C. (2007). Oxidative stress in the hippocampus, anxiety-like behavior and decreased locomotory and exploratory activity of adult rats: Effects of sub acute vitamin A supplementation at therapeutic doses. *Neurotoxicology*, *28*, 1191–1199.
Dell'Osso, B., Buoli, M., Baldwin, D. S., & Altamura, A. C. (2010). Serotonin norepinephrine reuptake inhibitors (SNRIs) in anxiety disorders: A comprehensive review of their clinical efficacy. *Human Psychopharmacology*, *25*, 17–29.
Di Loreto, S., Zimmitti, V., Sebastiani, P., Cervelli, C., Falone, S., & Amicarelli, F. (2008). Methylglyoxal causes strong weakening of detoxifying capacity and apoptotic cell death in rat hippocampal neurons. *The International Journal of Biochemistry & Cell Biology*, *40*, 245–257.
Dowlati, Y., Herrmann, N., Swardfager, W., Liu, H., Sham, L., Reim, E. K., et al. (2010). A meta-analysis of cytokines in major depression. *Biological Psychiatry*, *67*, 446–457.
Duman, R. S., Heninger, G. R., & Nestler, E. J. (1997). A molecular and cellular theory of depression. *Archives of General Psychiatry*, *54*, 597–606.
Einat, H., Yuan, P., Gould, T. D., Li, J., Du, J., Zhang, L., et al. (2003). The role of the extracellular signal-regulated kinase signaling pathway in mood modulation. *The Journal of Neuroscience*, *23*, 7311–7316.
Elenkov, I. J., Wilder, R. L., Chrousos, G. P., & Vizi, E. S. (2000). The sympathetic nerve—An integrative interface between two supersystems: The brain and the immune system. *Pharmacological Reviews*, *52*, 595–638.
Elks, C. M., Mariappan, N., Haque, M., Guggilam, A., Majid, D. S., & Francis, J. (2009). Chronic NF-{kappa}B blockade reduces cytosolic and mitochondrial oxidative stress and attenuates renal injury and hypertension in SHR. *American Journal of Physiology. Renal Physiology*, *296*, F298–F305.
Evans, D. L., Charney, D. S., Lewis, L., Golden, R. N., Gorman, J. M., Krishnan, K. R., et al. (2005). Mood disorders in the medically ill: Scientific review and recommendations. *Biological Psychiatry*, *58*, 175–189.
Fiore, M., Alleva, E., Probert, L., Kollias, G., Angelucci, F., & Aloe, L. (1998). Exploratory and displacement behavior in transgenic mice expressing high levels of brain TNF-alpha. *Physiology & Behavior*, *63*, 571–576.
Floyd, R. A. (1999a). Antioxidants, oxidative stress, and degenerative neurological disorders. *Proceedings of the Society for Experimental Biology and Medicine*, *222*, 236–245.
Floyd, R. A. (1999b). Neuroinflammatory processes are important in neurodegenerative diseases: An hypothesis to explain the increased formation of reactive oxygen and nitrogen species as major factors involved in neurodegenerative disease development. *Free Radical Biology & Medicine*, *26*, 1346–1355.
Floyd, R. A., Hensley, K., Jaffery, F., Maidt, L., Robinson, K., Pye, Q., et al. (1999). Increased oxidative stress brought on by pro-inflammatory cytokines in neurodegenerative processes and the protective role of nitrone-based free radical traps. *Life Sciences*, *65*, 1893–1899.
Gard, P. R., Haigh, S. J., Cambursano, P. T., & Warrington, C. A. (2001). Strain differences in the anxiolytic effects of losartan in the mouse. *Pharmacology, Biochemistry, and Behavior*, *69*, 35–40.
Gebicke-Haerter, P. J. (2001). Microglia in neurodegeneration: Molecular aspects. *Microscopy Research and Technique*, *54*, 47–58.

Ghosh, A., & Greenberg, M. E. (1995). Calcium signaling in neurons: Molecular mechanisms and cellular consequences. *Science*, *268*, 239–247.
Gingrich, J. A. (2005). Oxidative stress is the new stress. *Nature Medicine*, *11*, 1281–1282.
Giralt, M., Carrasco, J., Penkowa, M., Morcillo, M. A., Santamaría, J., Campbell, I. L., et al. (2001). Astrocyte-targeted expression of interleukin-3 and interferon-alpha causes region-specific changes in metallothionein expression in the brain. *Experimental Neurology*, *168*, 334–346.
Glaser, R., Rabin, B., Chesney, M., Cohen, S., & Natelson, B. (1999). Stress-induced immunomodulation: Implications for infectious diseases? *JAMA: The Journal of the American Medical Association*, *281*, 2268–2270.
Glass, C. K., Saijo, K., Winner, B., Marchetto, M. C., & Gage, F. H. (2010). Mechanisms underlying inflammation in neurodegeneration. *Cell*, *140*, 918–934.
Godbout, J. P., & Glaser, R. (2006). Stress-induced immune dysregulation: Implications for wound healing, infectious disease and cancer. *Journal of Neuroimmune Pharmacology*, *1*, 421–427.
Greco, T., & Fiskum, G. (2010). Neuroprotection through stimulation of mitochondrial antioxidant protein expression. *Journal of Alzheimer's Disease*, *20*(Suppl. 2), S427–S437.
Halliwell, B. (2006). Oxidative stress and neurodegeneration: Where are we now? *Journal of Neurochemistry*, *97*, 1634–1658.
Hermann, G. E., Rogers, R. C., Bresnahan, J. C., & Beattie, M. S. (2001). Tumor necrosis factor-alpha induces cFOS and strongly potentiates glutamate-mediated cell death in the rat spinal cord. *Neurobiology of Disease*, *8*, 590–599.
Hopkins, S. J., & Rothwell, N. J. (1995). Cytokines and the nervous system. I: Expression and recognition. *Trends in Neurosciences*, *18*, 83–88.
Hou, R., & Baldwin, D. S. (2012). A neuroimmunological perspective on anxiety disorders. *Human Psychopharmacology*, *27*, 6–14. doi: 10.1002/hup.1259. [Epub ahead of print].
Hovatta, L., Juhila, J., & Donner, J. (2011). Oxidative stress in anxiety and comorbid disorders. *Neuroscience Research*, *68*, 261–275.
Hovatta, L., Tennant, R. S., Helton, R., Marr, R. A., Singer, O., Redwine, J. M., et al. (2005). Glyoxalase 1 and glutathione reductase 1 regulate anxiety in mice. *Nature*, *438*, 662–666.
Howren, M. B., Lamkin, D. M., & Suls, J. (2009). Associations of depression with C-reactive protein, IL-1, and IL-6: A meta-analysis. *Psychosomatic Medicine*, *71*, 171–186.
Irwin, M., & Gillin, J. C. (1987). Impaired natural killer cell activity among depressed patients. *Psychiatry Research*, *20*, 181–182.
Jehn, C. F., Kuehnhardt, D., Bartholomae, A., Pfeiffer, S., Krebs, M., Regierer, A. C., et al. (2006). Biomarkers of depression in cancer patients. *Cancer*, *107*, 2723–2729.
Kaiser, F. C., Palmer, G. C., Wallace, A. V., Carr, R. D., Fraser-Rae, L., & Hallam, C. (1992). Antianxiety properties of the angiotensin II antagonist, DUP 753, in the rat using the elevated plus-maze. *Neuroreport*, *3*, 922–924.
Kassed, C. A., & Herkenham, M. (2004). NF-kappaB p50-deficient mice show reduced anxiety-like behaviors in tests of exploratory drive and anxiety. *Behavioural Brain Research*, *154*, 577–584.
Kim, M. J., Oh, S. J., Park, S. H., Kang, H. J., Won, M. H., Kang, T. C., et al. (2007). Neuronal loss in primary long-term cortical culture involves neurodegeneration-like cell death via calpain and p35 processing, but not developmental apoptosis or aging. *Experimental & Molecular Medicine*, *39*, 14–26.
Koh, K. B., & Lee, Y. (2004). Reduced anxiety level by therapeutic interventions and cell-mediated immunity in panic disorder patients. *Psychotherapy and Psychosomatics*, *73*, 286–292.
Kronfol, Z., & Remick, D. G. (1983). Impaired lymphocyte function in depressive illness. *Life Sciences*, *33*, 241–247.

Kronfol, Z., & Remick, D. G. (2000). Cytokines and the brain: Implications for clinical psychiatry. *The American Journal of Psychiatry, 157*, 683–694.
Kumar, A., Kumari, B., & Kumar, P. (2010). Protective effects of selective and non-selective cyclooxygenase inhibitors in an animal model of chronic stress. *Neuroscience Bulletin, 26*, 17–27.
Kumari, B., Kumar, A., & Dhir, A. (2007). Protective effect of non-selective and selective COX-2-inhibitors in acute immobilization stress-induced behavioral and biochemical alterations. *Pharmacological Reports, 59*, 699–707.
Kustova, Y., Sei, Y., Morse, H. C., Jr., & Basile, A. S. (1998). The influence of a targeted deletion of the IFNgamma gene on emotional behaviors. *Brain, Behavior, and Immunity, 12*, 308–324.
Lassmann, H., Brück, W., & Lucchinetti, C. (2001). Heterogeneity of multiple sclerosis pathogenesis: Implications for diagnosis and therapy. *Trends in Molecular Medicine, 7*, 115–121.
Leonard, B. E., & Myint, A. (2009). The psychoneuroimmunology of depression. *Human Psychopharmacology, 24*, 165–175.
Lesch, K. P. (2001). Mouse anxiety: The power of knockout. *The Pharmacogenomics Journal, 1*, 187–192.
Liou, H. C., & Baltimore, D. (1993). Regulation of the NF-kappa B/rel transcription factor and I kappa B inhibitor system. *Current Opinion in Cell Biology, 5*, 477–487.
Loftis, J. M., Huckans, M., & Morasco, B. J. (2010). Neuroimmune mechanisms of cytokine-induced depression: Current theories and novel treatment strategies. *Neurobiology of Disease, 37*, 519–533.
Longhi, L., Saatman, K. E., Raghupathi, R., Laurer, H. L., Lenz-linger, P. M., Riess, P., et al. (2001). A review and rationale for the use of genetically engineered animals in the study of traumatic brain injury. *Journal of Cerebral Blood Flow and Metabolism, 21*, 1241–1258.
Lonze, B. E., & Ginty, D. D. (2002). Function and regulation of CREB family transcription factors in the nervous system. *Neuron, 35*, 605–623.
Lopez, J. F., Akil, H., & Watson, S. J. (1999). Neural circuits mediating stress. *Biological Psychiatry, 46*, 1461–1471.
Lucas, S. M., Rothwell, N. J., & Gibson, R. M. (2006). The role of inflammation in CNS injury and disease. *British Journal of Pharmacology, 147*(Suppl 1), S232–S240.
Lutgendorf, S. K., Lamkin, D. M., DeGeest, K., Anderson, B., Dao, M., McGinn, S., et al. (2008). Depressed and anxious mood and T-cell cytokine expressing populations in ovarian cancer patients. *Brain, Behavior, and Immunity, 22*, 890–900.
Maccioni, R. B., Rojo, L. E., Fernandez, J. A., & Kuljis, R. O. (2009). The role of neuroimmunomodulation in Alzheimer's disease. *Annals of the New York Academy of Sciences, 1153*, 240–246.
Maes, M., Bosmans, E., Suy, E., Vandervorst, C., De Jonckheere, C., & Raus, J. (1990). Immune disturbances during major depression: Upregulated expression of interleukin-2 receptors. *Neuropsychobiology, 24*, 115–120.
Maes, M., Bosmans, E., Suy, E., Vandervorst, C., De Jonckheere, C., & Raus, J. (1991). Depression-related disturbances in mitogen-induced lymphocyte responses and interleukin-1 beta and soluble interleukin-2 receptor production. *Acta Psychiatrica Scandinavica, 84*, 379–386.
Maes, M., Kubera, M., Obuchowiczwa, E., Goehler, L., & Brzeszcz, J. (2011). Depression's multiple comorbidities explained by (neuro)inflammatory and oxidative & nitrosative stress pathways. *Neuroendocrinology Letters, 32*, 7–24.
Maes, M., Scharpe, S., Bosmans, E., Vandewoude, M., Suy, E., Uyttenbroeck, W., et al. (1992a). Disturbances in acute phase plasma proteins during melancholia: Additional evidence for the presence of an inflammatory process during that illness. *Progress in Neuro-Psychopharmacology & Biological Psychiatry, 16*, 501–515.

Maes, M., Scharpé, S., Meltzer, H. Y., Bosmans, E., Suy, E., Calabrese, J., et al. (1993). Relationships between interleukin-6 activity, acute phase proteins, and function of the hypothalamic-pituitary-adrenal axis in severe depression. *Psychiatry Research, 49*, 11–27.

Maes, M., Song, C., Lin, A., De Jongh, R., Van Gastel, A., Kenis, G., et al. (1992b). Leukocytosis, monocytosis and neutrophilia: Hallmarks of severe depression. *Journal of Psychiatric Research, 26*, 125–134.

Maes, M., Song, C., Lin, A., De Jongh, R., Van Gastel, A., Kenis, G., et al. (1998). The effects of psychological stress on humans: Increased production of pro-inflammatory cytokines and a Th1-like response in stress-induced anxiety. *Cytokine, 10*, 313–318.

Mantamadiotis, T., Lemberger, T., Bleckmann, S. C., Kern, H., Kretz, O., Villalba, Martin, et al. (2002). Disruption of CREB function in brain leads to neurodegeneration. *Nature Genetics, 31*, 47–54.

Martiney, J. A., Cuff, C., Litwak, M., Berman, J., & Brosnan, C. F. (1998). Cytokine-induced inflammation in the central nervous system revisited. *Neurochemical Research, 23*, 349–359.

Martinowich, K., Manji, H., & Lu, B. (2007). New insights into BDNF function in depression and anxiety. *Nature Neuroscience, 10*, 1089–1093.

Masood, A., Nadeem, A., Mustafa, S. J., & O'Donnell, J. M. (2008). Reversal of oxidative stress-induced anxiety by inhibition of phosphodiesterase-2 in mice. *The Journal of Pharmacology and Experimental Therapeutics, 326*, 369–379.

Mattson, M. P., & Camandola, S. (2001). NF-kappaB in neuronal plasticity and neurodegenerative disorders. *The Journal of Clinical Investigation, 107*, 247–254.

McEwen, B. S. (2004). Structural plasticity of the adult brain: How animal models help us understand brain changes in depression and systemic disorders related to depression. *Dialogues in Clinical Neuroscience, 6*, 119–133.

McEwen, B. S., & Wingfield, J. C. (2003). The concept of allostasis in biology and biomedicine. *Hormones and Behavior, 43*, 2–15.

Menard, J., & Treit, D. (1999). Effects of centrally administered anxiolytic compounds in animal models of anxiety. *Neuroscience and Biobehavioral Reviews, 23*, 591–613.

Merrill, J. E., & Benveniste, E. N. (1996). Cytokines in inflammatory brain lesions: Helpful and harmful. *Trends in Neurosciences, 19*, 331–338.

Meyers, C. A., Albitar, M., & Estey, E. (2005). Cognitive impairment, fatigue, and cytokine levels in patients with acute myelogenous leukemia or myelodysplastic syndrome. *Cancer, 104*, 788–793.

Mhatre, M., Floyd, R. A., & Hensley, K. (2004). Oxidative stress and neuroinflammation in Alzheimer's disease and amyotrophic lateral sclerosis: Common links and potential therapeutic targets. *Journal of Alzheimer's Disease, 6*, 147–157.

Miller, A. H., Maletic, V., & Raison, C. L. (2009). Inflammation and its discontents: The role of cytokines in the pathophysiology of major depression. *Biological Psychiatry, 65*, 732–741.

Munck, A., & Guyre, P. M. (1991). Glucocorticoids and immune function. In R. Ader, D. L. Felten & N. Cohen (Eds.), *Psychoneuroimmunology* (pp. 447–474). San Diego, CA: Academic Press.

Musselman, D. L., Lawson, D. H., Gumnick, J. F., Manatunga, A. K., Penna, S., Goodkin, R. S., et al. (2001). Paroxetine for the prevention of depression induced by high-dose interferon alfa. *The New England Journal of Medicine, 344*, 961–966.

Neumann, H. (2000). The immunological microenvironment in the CNS: Implications on neuronal cell death and survival. *Journal of Neural Transmission. Supplementum, 59*, 59–68.

Ng, F., Berk, M., Dean, O., & Bush, A. I. (2008). Oxidative stress in psychiatric disorders: Evidence base and therapeutic implications. *The International Journal of Neuropsychopharmacology, 11*, 851–876.

Nguyen, M. D., Julien, J. P., & Rivest, S. (2002). Innate immunity: The missing link in neuroprotection and neurodegeneration? *Nature Reviews. Neuroscience*, *3*, 216–227.

O'Donovan, A., Hughes, B. M., Slavich, G. M., Lynch, L., Cronin, M. T., O'Farrelly, C., et al. (2010). Clinical anxiety, cortisol and interleukin-6: Evidence for specificity in emotion-biology relationships. *Brain, Behavior, and Immunity*, *24*, 1074–1077.

Pandey, S. C. (2003). Anxiety and alcohol abuse disorders: A common role for CREB and its target, the neuropeptide Y gene. *Trends in Pharmacological Sciences*, *24*, 456–460.

Perry, V. H., Bolton, S. J., Anthony, D. C., & Betmouni, S. (1998). The contribution of inflammation to acute and chronic neurodegeneration. *Research in Immunology*, *149*, 721–725.

Peruga, I., Hartwig, S., Thöne, J., Hovemann, B., Gold, R., Juckel, G., et al. (2011). Inflammation modulates anxiety in an animal model of multiple sclerosis. *Behavioural Brain Research*, *220*, 20–29.

Pitsavos, C., Panagiotakos, D. B., Papageorgiou, C., Tsetsekou, E., Soldatos, C., & Stefanadis, C. (2006). Anxiety in relation to inflammation and coagulation markers, among healthy adults: The ATTICA study. *Atherosclerosis*, *185*, 320–326.

Probert, L., Akassoglou, K., Pasparakis, M., Kontogeorgos, G., & Kollias, G. (1995). Spontaneous inflammatory demyelinating disease in transgenic mice showing central nervous system-specific expression of tumor necrosis factor alpha. *Proceedings of the National Academy of Sciences of the United States of America*, *92*, 11294–11298.

Raha, S., & Robinson, B. H. (2001). Mitochondria, oxygen free radicals, and apoptosis. *American Journal of Medical Genetics*, *106*, 62–70.

Raison, C. L., Borisov, A. S., Broadwell, S. D., Capuron, L., Woolwine, B. J., Jacobson, I. M., et al. (2005). Depression during pegylated interferon-alpha plus ribavirin therapy: Prevalence and prediction. *The Journal of Clinical Psychiatry*, *66*, 41–48.

Raison, C. L., Capuron, L., & Miller, A. H. (2006). Cytokines sing the blues: Inflammation and the pathogenesis of depression. *Trends in Immunology*, *27*, 24–31.

Rammal, H., Bouayed, J., Younos, C., & Soulimani, R. (2008). Evidence that oxidative stress is linked to anxiety-related behaviour in mice. *Brain, Behavior, and Immunity*, *22*, 1156–1159.

Ransohoff, R. M., & Perry, V. H. (2009). Microglial physiology: Unique stimuli, specialized responses. *Annual Review of Immunology*, *27*, 119–145.

Reichenberg, A., Yirmiya, R., Schuld, A., Kraus, T., Haack, M., Morag, A., et al. (2001). Cytokine-associated emotional and cognitive disturbances in humans. *Archives of General Psychiatry*, *58*, 445–452.

Rezai-Zadeh, K., Gate, D., & Town, T. (2009). CNS infiltration of peripheral immune cells: D-Day for neurodegenerative disease? *Journal of Neuroimmune Pharmacology*, *4*, 462–475.

Saavedra, J. M., Armando, I., Bregonzio, C., Juorio, A., Macova, M., Pavel, J., et al. (2006). A centrally acting, anxiolytic angiotensin II AT1 receptor antagonist prevents the isolation stress-induced decrease in cortical CRF1 receptor and benzodiazepine binding. *Neuropsychopharmacology*, *31*, 1123–1134.

Saavedra, J. M., Sánchez-Lemus, E., & Benicky, J. (2011). Blockade of brain angiotensin II AT1 receptors ameliorates stress, anxiety, brain inflammation and ischemia: Therapeutic implications. *Psychoneuroendocrinology*, *36*, 1–18.

Sakic, B., Szechtman, H., Talangbayan, H., Denburg, S. D., Carbotte, R. M., & Denburg, J. A. (1994). Disturbed emotionality in autoimmune MRL-lpr mice. *Physiology & Behavior*, *56*, 609–617.

Salim, S., Asghar, M., Chugh, G., Taneja, M., Xia, Z., & Saha, K. (2010a). Oxidative stress: A potential recipe for anxiety, hypertension and insulin resistance. *Brain Research*, *1359*, 178–185.

Salim, S., Asghar, M., Taneja, M., Hovatta, I., Chugh, G., Vollert, C., et al. (2011). Potential contribution of oxidative stress and inflammation to anxiety and hypertension. *Brain Research*, *1404*, 63–71.

Salim, S., Sarraj, N., Taneja, M., Saha, K., Tejada-Simon, M. V., & Chugh, C. (2010b). Moderate treadmill exercise prevents oxidative stress-induced anxiety-like behavior in rats. *Behavioural Brain Research*, *208*, 545–552.
Sapolsky, R. M., Romero, L. M., & Munck, A. U. (2000). How do glucocorticoids influence stress responses? Integrating permissive, suppressive, stimulatory, and preparative actions. *Endocrine Reviews*, *21*, 55–89.
Savoia, C., & Schiffrin, E. L. (2007). Vascular inflammation in hypertension and diabetes: Molecular mechanisms and therapeutic interventions. *Clinical Science (London, England)*, *112*, 375–384.
Sayre, L. M., Perry, G., & Smith, M. A. (2008). Oxidative stress and neurotoxicity. *Chemical Research in Toxicology*, *21*, 172–188.
Schaaf, M. J., Workel, J. O., Vreugdenhil, E., Oitzl, M. S., & de Kloet, E. R. (2001). Correlation between hippocampal BDNF mRNA expression and memory performance in senescent rats. *Brain Research*, *915*, 227–233.
Schneiderman, N., Ironson, G., & Siegel, S. D. (2005). Stress and health: Psychological, behavioral, and biological determinants. *Annual Review of Clinical Psychology*, *1*, 607–628.
Schrott, L. M., & Crnic, L. S. (1996). Increased anxiety behaviors in autoimmune mice. *Behavioral Neuroscience*, *110*, 492–502.
Shekhar, A., Johnson, P. L., Sajdyk, T. J., Fitz, S. D., Keim, S. R., Kelley, P. E., et al. (2006). Angiotensin-II is a putative neurotransmitter in lactate-induced panic-like responses in rats with disruption of GABAergic inhibition in the dorsomedial hypothalamus. *The Journal of Neuroscience*, *26*, 9205–9215.
Shin, L. M., Rauch, S. L., & Pitman, R. K. (2006). Amygdala, medial prefrontal cortex, and hippocampal function in PTSD. *Annals of the New York Academy of Sciences*, *1071*, 67–79.
Shumway, S. D., Maki, M., & Miyamoto, S. (1999). The PEST domain of IkappaBalpha is necessary and sufficient for in vitro degradation by mu-calpain. *The Journal of Biological Chemistry*, *274*, 30874–30881.
Souza, C. G., Moreira, J. D., Siqueira, I. R., Pereira, A. G., Rieger, D. K., Souza, D. O., et al. (2007). Highly palatable diet consumption increases protein oxidation in rat frontal cortex and anxiety-like behavior. *Life Sciences*, *81*, 198–203.
Sternberg, E. M. (1997). Neural-immune interactions in health and disease. *The Journal of Clinical Investigation*, *100*, 2641–2647.
Stoll, G., Jander, S., & Schroeter, M. (2000). Cytokines in CNS disorders: Neurotoxicity versus neuroprotection. *Journal of Neural Transmission. Supplementum*, *59*, 81–89.
Streit, W. J., Walter, S. A., & Pennell, N. A. (1999). Reactive microgliosis. *Progress in Neurobiology*, *57*, 563–581.
Takkouche, B., Regueira, C., & Gestal-Otero, J. J. (2001). A cohort study of stress and the common cold. *Epidemiology*, *12*, 345–349.
Thornalley, P. J. (2003). Glyoxalase I—Structure, function and a critical role in the enzymatic defence against glycation. *Biochemical Society Transactions*, *31*, 1343–1348.
Thornalley, P. J. (2005). Dicarbonyl intermediates in the maillard reaction. *Annals of the New York Academy of Sciences*, *1043*, 111–117.
Tremper-Wells, B., & Vallano, M. L. (2005). Nuclear calpain regulates Ca2+-dependent signaling via proteolysis of nuclear Ca2+/calmodulin-dependent protein kinase type IV in cultured neurons. *The Journal of Biological Chemistry*, *280*, 2165–2175.
Viviani, B., Bartesaghi, S., Corsini, E., Galli, C. L., & Marinovich, M. (2004). Cytokines role in neurodegenerative events. *Toxicology Letters*, *149*, 85–89.
Viviani, B., & Boraso, M. (2011). Cytokines and neuronal channels: A molecular basis for age-related decline of neuronal function? *Experimental Gerontology*, *46*, 199–206.
Viviani, B., Gardoni, F., & Marinovich, M. (2007). Cytokines and neuronal ion channels in health and disease. *International Review of Neurobiology*, *82*, 247–263.

Wakita, T., Shintani, F., Yagi, G., Asai, M., & Nozawa, S. (2001). Combination of inflammatory cytokines increases nitrite and nitrate levels in the paraventricular nucleus of conscious rats. *Brain Research, 905*, 12–20.
Xiao, B. G., & Link, H. (1998). Immune regulation within the central nervous system. *Journal of the Neurological Sciences, 157*, 1–12.
Zhou, F. L., Zhang, W. G., Wei, Y. C., Xu, K. L., Hui, L. Y., Wang, X. S., et al. (2005). Impact of comorbid anxiety and depression on quality of life and cellular immunity changes in patients with digestive tract cancers. *World Journal of Gastroenterology 11*, 2313–2318.
Zorrilla, E. P., Luborsky, L., McKay, J. R., Rosenthal, R., Houldin, A., Tax, A., et al. (2001). The relationship of depression and stressors to immunological assays: A meta-analytic review. *Brain, Behavior, and Immunity, 15*, 199–226.
Zorrilla, E. P., Redei, E., & DeRubeis, R. J. (1994). Reduced cytokine levels and T-cell function in healthy males: Relation to individual differences in subclinical anxiety. *Brain, Behavior, and Immunity, 8*, 293–312.

CHAPTER TWO

Inflammation-Related Disorders in the Tryptophan Catabolite Pathway in Depression and Somatization

George Anderson*, Michael Maes[†1], Michael Berk[‡§¶||]
*CRC, Glasgow, United Kingdom
[†]Piyavate Hospital, Bangkok, Thailand
[‡]School of Medicine, Deakin University, Melbourne Australia
[§]Orygen Youth Health Research Centre, Centre for Youth Mental Health, Parkville, Victoria, Australia
[¶]The Mental Health Research Institute of Victoria, Parkville, Victoria, Australia
[||]Department of Psychiatry, Melbourne University, Parkville, Victoria, Australia
[1]Corresponding author: e-mail address: dr.michaelmaes@hotmail.com

Contents

Abstract

A recent study—comparing those with depression, somatization, comorbid depression + somatization, and controls—showed specific changes in the tryptophan catabolite (TRYCAT) pathway in somatization, specifically lowered tryptophan and kynurenic acid, and increased kynurenine/kynurenic acid (KY/KA) and kynurenine/tryptophan ratios. These findings suggest that somatization and depression with somatization are characterized by increased activity of indoleamine 2,3-dioxygenase and disorders in kynurenine aminotransferase activity, which carry a neurotoxic potential.

This chapter reviews the evidence that the TRYCAT pathway may play a pathophysiological role in the onset of somatization and depression with somatization and, furthermore, suggests treatment options based on identified pathophysiological processes.

Lowered plasma tryptophan may be associated with enhanced pain, autonomic nervous system responses, gut motility, peripheral nerve function, ventilation, and cardiac dysfunctions. The imbalance in the KY/KA ratio may increase pain, intestinal

Advances in Protein Chemistry and Structural Biology, Volume 88
ISSN 1876-1623
http://dx.doi.org/10.1016/B978-0-12-398314-5.00002-7

hypermotility, and peripheral neuropathy through effects of KY and KA acid, both centrally and peripherally, at the *N*-methyl-D-aspartate receptor (NMDAR), G-protein-coupled receptor-35 (GPR35), and aryl hydrocarbon receptor (AHr). These alterations in the TRYCAT pathway in somatization and depression may interface with the role of the mu-opioid, serotonin, and oxytocin systems in the regulation of stress reactions and early attachment.

It is hypothesized that irregular parenting and insecure attachment paralleled by chronic stress play a key role in the expression of variations in the TRYCAT pathway—both centrally and peripherally—driving the etiology of somatization through interactions with the mu-opioid receptors. Therefore, the TRYCAT pathway, NMDARs, GPR35, and AHrs may be new drug targets in somatization and depression with somatizing. We lastly review new pathophysiologically driven drug candidates for somatization, including St. John's wort, resveratrol, melatonin, agomelatine, *Garcinia mangostana* (γ-mangostin), *N*-acetyl cysteine, and pamoic acid.

1. INTRODUCTION

There is a high comorbidity between depression and somatization. For example, depression is the most common comorbid diagnosis of somatization in a primary care setting (Brown, Golding, & Smith, 1990). Recently, we showed that somatic symptoms are a major feature of depression and predict chronicity and severity of depression (Maes, 2009). Somatic symptoms that frequently occur in depression are aches and pain, muscular tension, fatigue, concentration difficulties, failing memory, irritability, irritable bowel, headache, and malaise (Maes, 2009).

The presentation by patients with symptoms of no overt medical cause poses a major problem in medicine, in terms of both classification and treatment. Between 20% and 60% of primary care patients present with physical symptoms that have no medical basis or are discordant with the degree of illness indicated by objective tests or observable signs (Fink, Sorensen, Engberg, Holm, & Munk-Jorgensen, 1999). The diagnostic and statistical manual of mental disorders, version IV suggests the diagnosis of "somatization disorder" for patients with at least eight somatoform symptoms from four body sites. Confound to the classification and treatment of somatization disorder is its common comorbidity with depression and mood disorders (Koh, Kim, Kim, & Park, 2005).

Within psychotherapy, somatization is often seen an alternative mode for the expression of depression and distress. Much of the work in this area has been driven by theoretical concepts of attachment, focusing on the early work of Bowlby (1969). The concept of attachment refers to a motivational behavioral system that is activated when an individual feels threatened. The attachment system directs goal-connected behavior. The function of which is to maintain feelings of security (Sroufe, 1995). The individual forms expectations of attachment experiences, on the basis of earlier experiences. Substantial research links the quality of attachment to mental health problems (Dozier, Stovall-McClough, & Albus, 2008; Surcinelli, Rossi, Montebarocci, & Baldaro, 2010), including somatization. Somatization can be mediated by classical conditioning stimuli and expectation (Dignam, Parry, & Berk, 2010). In the context of medical treatment, somatization can manifest as a nocebo reaction, the counterpoint of the placebo effect (Data-Franco & Berk, 2012). Individuals who have insecure models of attachment to significant others report higher levels of somatic symptoms (Noyes et al., 2003; Wearden, Lamberton, Crook, & Walsh, 2005). However, the mechanisms by which insecure attachment might be linked to somatization are poorly understood. Anger proneness in males and anger suppression in females increase somatization (Liu, Cohen, Schulz, & Waldinger, 2011), suggesting that alterations in how negative emotions are dealt with act as an intermediary. Interestingly, single nucleotide polymorphisms of the mu-opioid receptor modulate levels of attachment in many different species (Shayit, Nowak, Keller, & Weller, 2003; Warnick, McCurdy, & Sufka, 2005), as does oxytocin (Strathearn, 2011). Morphine, including via central mu-opioid receptors, decreases levels of oxytocin release (Ortiz-Miranda, Dayanithi, Custer, Treistman, & Lemos, 2005), suggesting that variation in mu-opioid receptor activity will alter attachment, as well as the well-known social affiliation behaviors associated with oxytocin. Stress-induced cortisol, both directly and indirectly, increases levels of the mu-opioid receptor and astrocyte preproenkephalin (Chong et al., 2006; Ruzicka & Akil, 1997). Alterations in the opioidergic system in conjunction with nonresponsive parenting may be an early developmental etiological factor in insecure attachment. Such opioidergic data have links to recent data suggesting the importance of altered immune cell responses within the CNS (Andreica-Sandica, Panaete, Pascanu, Sarban, & Andreica, 2011; Euteneuer et al., 2011) and to data emphasizing the importance of altered activity of the tryptophan catabolite (TRYCAT) pathway (Maes, Galecki, Verkerk, & Rief, 2011; Maes & Rief, 2012). Perinatal exposure to serotonin agonists decreases social affiliation and increases anxiety (Martin, Liu, & Wang, 2012), highlighting the importance of the early developmental period in the modulation of later social and affective processing.

2. TRYPTOPHAN AND THE TRYCAT PATHWAY

The induction of indoleamine 2,3-dioxygenase (IDO) or tryptophan 2,3-dioxygenase (TDO) with consequent synthesis of TRYCATs depletes tryptophan and plays a role in the onset of depression (Maes, Leonard, Myint, Kubera, & Verkerk, 2011). The conversion of tryptophan to TRYCATs takes tryptophan away from serotonin and melatonin synthesis. Kynurenine (KY) is the first and rate-limiting step in the TRYCAT pathway. KY is further metabolized into kynurenic acid (KA) by kynurenine 2,3-aminotransferase (KAT). In TDO-expressing astrocytes and neurons, this is usually the endpoint of the TRYCAT pathway. In IDO-expressing cells, KA is further catabolized to quinolinic acid (QA) and eventually to NAD^+ (nicotinamide). IDO is widely expressed in human tissues, including in brain microglia, in peripheral organs, and in immune cells (Niimi, Nakamura, Nawa, & Ichihara, 1983). IDO is powerfully induced not only by interferon-γ (IFNγ) but also by the proinflammatory cytokines, interleukin-1β (IL-1β), tumor necrosis factor-α (TNFα), and IL-18 (Liebau et al., 2002; Oxenkrug, 2007). TDO is highly expressed in the liver, astrocytes, and some neurons (Ohira et al., 2010; Ren & Correia, 2000). TDO is primarily induced by cortisol (Ren & Correia, 2000).

Somatization is accompanied by signs of monocytic activation, for example, increased levels of soluble IL-1 receptor antagonist (Rief et al., 2001). Depression is accompanied by cell-mediated immune activation, as indicated by increased IFNγ, neopterin (Euteneuer et al., 2011), IL-1, IL-18, and TNFα (Maes, 1995; Maes, Galecki, et al., 2011; Prossin et al., 2011) all being associated with IDO induction. Both depression and somatization are associated with increases in cortisol and other signs of hypothalamic–pituitary–adrenal axis activation, driving TDO activation (Rief, Shaw, & Fichter, 1998; Unschuld et al., 2010). The importance of TDO to serotonin is highlighted by the 20-fold increase in brain serotonin in TDO KO rodents (Funakoshi, Kanai, & Nakamura, 2011). However, increased TRYCAT pathway activation does not simply increase depression via decreased serotonin. KY has depressogenic, anxiogenic, excitotoxic, and neurotoxic effects, whereas KA has neuroprotective and antinociceptic effects (Cosi et al., 2011). Therefore, the KY/KA ratio indicates not only KAT activity but also the neurotoxic and nociceptive potential generated during TDO and IDO activation. Increased QA in microglia of the anterior cingulate has recently been shown in depressed suicide patients (Steiner et al., 2011).

3. THE TRYCAT PATHWAY IN SOMATIZATION

In a study comparing depression, somatization, comorbid depression + somatization, and controls, the ratios of KY/KA and KY/tryptophan were significantly increased in the somatization group (Maes, Galecki, et al., 2011). Importantly, both ratios were positively correlated with the severity of somatization, but not with depression. In this study, tryptophan levels were negatively correlated with somatization, but not depression. Also KY and KA were significantly correlated in the three other groups, but not in somatization, suggesting changes in the regulation of KAT. This indicates that alterations in the TRYCAT pathway, classically seen as a hallmark of depression, may in fact be more germane to somatization. It remains to be determined whether central and/or peripheral increases in KY are more important. Gender differences in depression, and, in particular, in somatization, could then be mediated by increased responsiveness of IDO in females (Bonaccorso et al., 2002; Maes, Leonard, et al., 2011). Comparisons with previous studies looking at TRYCAT pathway changes in depression are confounded by the fact that no studies previously controlled for somatization. Increased KY/KA ratio suggests important changes in the level or regulation of KAT (Laugeray et al., 2010), or perhaps in levels of KAT isoforms (I–IV, with IV exclusively expressed in mitochondria; Han, Robinson, Cai, Tagle, & Li, 2011). KAT-III alleles are associated with depression, suggesting a role for KAT-III in altered KY/KA ratio (Claes et al., 2011). However, this study did not look at levels of somatization.

4. ACTIVATION OF THE TRYCAT PATHWAY MAY CAUSE SOMATIZATION

Depletion of plasma tryptophan through induction of the TRYCAT pathway may play a role in the generation of somatic symptoms. Alterations in the availability of plasma tryptophan determine brain serotonin synthesis (Moir & Eccleston, 1968). Depletion of plasma tryptophan is associated with depressive symptoms in some patients who had previously suffered from depression (Maes & Meltzer, 1995). Depletion of plasma tryptophan can be modeled through administration of tryptophan-free drinks that contain large concentrations of competing amino acids, that is, amino acids that compete for the same cerebral amino acid transporter. Tryptophan depletion techniques have been shown to cause somatization in humans, such as enhanced physiological responses to stress; increased visceral perception, pain, and urge scores during rectal distention in patients with irritable bowel syndrome; enhanced basal ventilation; increased headache, light-induced pain and photophobia, and nausea in patients with migraine; decreased heart rate variability, lowered heart rate in attention deficit disorder, and enhanced blood pressure in response to stressors; and enhanced autonomic stress responses in social anxiety and increased impulsivity (Booij et al., 2006; Davies et al., 2006; Dougherty, Richard, James, & Mathias, 2010; Drummond, 2006; Kilkens, Honig, van Nieuwenhoven, Riedel, & Brummer, 2004; Struzik, Duffin, Vermani, Hegadoren, & Katzman, 2002; van Veen et al., 2009; Zepf, Holtmann, Stadler, Wöckel, & Poustka, 2009).

KY increases pain and gut motility (Stone & Darlington, 2002), whereas KA is antinociceptive, inhibiting intestinal hypermotility (Kaszaki et al., 2008). Elevated KY is associated with peripheral neuropathy (Huengsberg, Winer, Round, Gompels, & Shahmanesh, 1998), whereas KA, via glutamate and *N*-methyl-D-aspartate receptor (NMDAR), is antinociceptive (Cairns et al., 2003). Peripheral NMDARs in deep tissues are involved in deep tissue pain (Cairns et al., 2003). For example, glutamate injection in the masseter muscle provokes afferent discharges in rats and muscle pain in humans through activation of peripheral NMDARs (Cairns et al., 2003). Peripheral NMDARs are targets for the treatment of neuropathic pain (Wu & Zhuo, 2009). NMDARs expressed on spinal afferent neurons are upregulated in the lumbosacral dorsal root ganglia (DRG) following experimental colitis (Li et al., 2006). Peripheral NMDARs play a role in behavioral pain responses to colonic distention, suggesting that these receptors are important in visceral pain transmission (McRoberts et al., 2001).

TRYCATs regulate the glutamate-induced activation of the NMDARs. KA is the only known endogenous antagonist of NMDARs, although KA only directly antagonizes the glycine site of the NMDAR at higher than physiological concentrations. However, KA acts as an inhibitor of glutamate release (Németh, Toldi, & Vécsei, 2005). Thus, a lowered KAT activity or KA levels—as detected in somatization—may be involved in the maintenance of persistent pain-related behaviors. Importantly, peripheral inflammation, as detected in somatization (Klengel et al., 2011; Rief et al., 2001), increases the expression of peripheral NMDARs, leading to behavioral sensitization during inflammatory pain (Yang, Yang, Xie, Liu, & Hu, 2009). KA is also involved in the control of cardiovascular function by acting at the rostral ventrolateral medulla (rVLM) in the

CNS (Colombari et al., 2001). Spontaneously, hypertensive rats, the most widely used animal model for studying genetic hypertension, have abnormally low KA levels in the rVLM. So changes in the KY/KA ratio may modulate nociception and wider medical conditions, driving increased multiple area symptoms reporting as found in somatization.

The G-protein-coupled receptor-35 (GPR35) is known to mediate powerful antinociceptive effects and is expressed both centrally and peripherally. It elicits calcium mobilization and inositol phosphate production via G(qi/o) proteins. The GPR35 is highly expressed in leukocytes and the gastrointestinal tract, but also more widely, including in the CNS (Wang et al., 2006). KA is an endogenous ligand for the GPR35 (Wang et al., 2006), inducing antinociception (Cosi et al., 2011). Decreased KA in somatization will therefore impact on nociception via the modulation of both the NMDAR and the GPR35. Stimulation of GPR35 in rat sympathetic neurons inhibits N-type calcium channels, suggesting a potential role for GPR35 in regulating neuronal excitability and transmitter release (Guo, Williams, Puhl, & Ikeda, 2008). KA activation of the GPR35 modulates adhesion of leukocytes to vascular endothelium, suggestive of a role in inflammatory states (Barth et al., 2009). The GPR35 is expressed in glia, where its activation by KA, via Gi coupling, decreases levels of cAMP and PKA (Cosi et al., 2011). This may suggest a negative feedback on KA production, which is enhanced in astrocytes by cAMP activation (Luchowska et al., 2009).

KA, at nanomolar concentrations, inhibits heat shock-induced FGF1 from nonneuronal cells (Di Serio et al., 2005). Neurotrophin release in damaged or inflamed tissues is an important event in nociceptor activation (Pezet & McMahon, 2006), and the inhibition of neurotrophin release may contribute to the potent analgesic action of KA. The expression of the GPR35 in DRG is another site for KA analgesic actions. Pamoic acid, a newly identified GPR35 agonist, also has significant antinociceptive actions in inflammation-induced pain (Zhao et al., 2010). More in depth studies are needed to address whether the anti-inflammatory effects of pamoic acid and KA are mediated by GPR35 activation centrally or peripherally.

5. THE TRYCAT PATHWAY, THE CNS, AND SOMATIZATION

In contrast to KY, KA has only a very limited ability to cross the blood–brain barrier (BBB) (Németh et al., 2005). Thus, the KY pool in the brain is elevated by crossing the BBB in situations where peripheral

IDO is induced, causing a neurotoxic environment in the brain (Guillemin et al., 2001). This could result in increased formation of QA and therefore increased proinflammatory responses, neuronal apoptosis and neuroexcitatory, neurotoxic and neurodegenerative effects (Braidy, Grant, Adams, Brew, & Guillemin, 2009; Guillemin et al., 2000; Maes, Mihaylova, Ruyter, Kubera, & Bosmans, 2007). Of course, the findings on lowered plasma KA levels in somatization cannot be translated into the brain. But if the KAT enzyme activity were decreased, it could affect brain functions. KY has anxiogenic effects in both animals (Lapin, 1996; Vécsei & Beal, 1990) and humans (Orlikov, Prakh'e, & Ryzhov, 1990; Orlikov & Ryzov, 1991). KA, on the other hand, has an anxiolytic pharmacological profile (Lapin, 1998; Schmitt, Graeff, & Carobrez, 1990). Since KA, at high concentration, is the only endogenous antagonist of NMDARs, lowered KA may result in NMDA activation, in, for example, inflammatory conditions, which, in turn, may result in excitotoxic neuronal cell loss (Sapko et al., 2006; Swartz, During, Freese, & Beal, 1990). KA also antagonizes the α-amino-3-hydroxyl-5-methyl-4-isoxazole-propionate and kainate receptors (Hilmas et al., 2001), modulates the expression of α4β2 nicotinic acetylcholine receptors (nAChR), and acts as an antagonist of α7-nAChR (Han, Cai, Tagle, & Li, 2010; Németh et al., 2005). α4β2-nAChRs are implicated in perception, cognition, and emotion (Picciotto et al., 1995), whereas α7-nAChRs modulate cortex arousal and cognitive processes (Boess et al., 2007).

It is widely believed that stress-induced cortisol increases astrocyte and neuronal TDO. A genetic, epigenetic, and/or local environment inhibitory regulation of KAT-II would increase the KY/KA ratio. An increase in KY may activate the aryl hydrocarbon receptor (AHr) in microglia, as has recently been shown in the context of brain tumors (Opitz et al., 2011). The activation of the AHr increases IDO in many peripheral cells, and the effects of KY and KA need to be tested in microglia. AHr, cytokine, or IFNγ-induced IDO and QA in peripheral cells and perhaps microglia increase NMDAR excitatory activity, with excitotoxicity occurring at higher concentrations. However, this may be significantly modulated by variations in picolinic acid (PA), another TRYCAT, usually coinduced with QA. PA, via zinc chelation, prevents the excitotoxic effects of QA, allowing excitatory effects. Released microglia QA will also modulate astrocyte responses, including increasing IL-1β (Ting, Brew, & Guillemin, 2009). If astrocyte IL-1β were indeed increased via inflammasome induction, then PA would prevent this. Zinc chelators, like PA, inhibit pannexin-1 in astrocytes, preventing astroycte inflammasome induction (Silverman et al., 2009).

PA, and the PA/QA ratio, may therefore play a crucial role in glia interactions that are crucial to the inflammatory response, neuronal activity, and blood–brain barrier permeability (BBBp). Mu-opioid receptor activation is an early event in increasing BBBp (Ting, Cushenberry, Friedman, & Loh, 1997), and this may be relevant to the influence of peripheral immune cells and inflammatory factors on processing at specific CNS sites.

Inflammatory mediators positively correlate with baseline regional mu-opioid receptor binding potential and with sadness-induced mu-opioid system activation in the subgenual anterior cingulate, ventral basal ganglia, and amygdala (Prossin et al., 2011). As well as a role in the regulation of attachment, the mu-opioid receptor inhibits the cAMP/PKA pathway (Talbot et al., 2010). The inhibition of the cAMP pathway would decrease levels of astrocyte and neuronal TDO induction, with concurrent impacts on circadian gene regulation (Luchowska et al., 2009; Zhang et al., 2011). It is unknown if circadian genes have any differential impact on the TRYCAT pathway, including KAT-II, and the ratios of KY/KA and KY/tryptophan. This could suggest a role for the mu-opioid receptors, via Gi and cAMP inhibition, in the coordinated regulation of circadian genes and the TRYCAT pathway. This is important to investigate, as it links to data in mood disorders showing changes in mu-opioid receptor, circadian genes, and the TRYCAT pathway (Luchowska et al., 2009; Prossin et al., 2011; Soria et al., 2010). Heightened KY would increase neurotoxicity and possibly increase microglia IDO and QA via AHr activation. The AHr is powerfully regulated by a circadian rhythm (Mukai, Lin, Peterson, Cooke, & Tischkau, 2008; Mukai & Tischkau, 2007), suggesting circadian changes in glia will impact on levels of IDO, QA, and PA in microglia. Stress, through increased cortisol, may increase mu-opioid receptor transcription, and this may further increase the KY/KA ratio. As to whether early nonresponsive parenting, perhaps paralleling chronic unpredictable mild stress (CUMS), drives alterations in inflammatory mediators, mu-opioid receptor, and TRYCAT pathway remains to be fully investigated, including its relevance to the early etiology of, and ongoing changes in, somatization.

The above suggests some early developmental parallels to the effects of CUMS. CUMS decreases KA and increases neurotoxic TRYCATs, for example, QA, in the amygdala (Laugeray et al., 2011), and may explain the mechanisms mediating increased QA in microglia in the subgenual anterior cingulate in a depressed suicide sample (Steiner et al., 2011). Is irregular parenting and insecure attachment paralleled by CUMS, suggesting a crucial role for variations in the TRYCAT pathway, driving the etiology of somatization? If so, changes in the amygdala may be crucial. The amygdala may be a key site as it matures earlier than other brain regions and has a powerful developmental influence on the cortex and on motivational outputs via the nucleus accumbens/ventral tegmental area (NA/VTA) junction (Anderson, 2011; McGinty & Grace, 2009a; Salm et al., 2004). This may be relevant to conceptualizations of attachment as a means of regulating goal-connected behaviors. Layer V of the prefrontal cortex (PFC) and anterior cingulate is the major output layer and is the layer with most alterations in mood disorders. After very early projections into all cortex layers, the amygdala projections retract, leaving inputs only into layers II and V of the PFC. Are the heightened mu-opioid receptor, and QA/KA and KY/KA ratios in the amygdala interacting with, or driving, similar changes in the anterior cingulate, altering the way emotion is processed and setting patterns for interpersonal affective interactions? The amygdala is able to override the influence of the cortex and hippocampus on motivated behavioral outputs from the NA/VTA junction (McGinty & Grace, 2009b), suggesting that early developmental biases in amygdala growth and activity would coordinate changes in the anterior cingulate with enhanced influence of the amygdala on NA/VTA motivated behavioral outputs. This suggests an early development-driven change in how emotions are processed, linking insecure attachment with later psychiatric disorders, including somatization. The amygdala and the anterior cingulate are also important sites for pain regulation. Alterations in central emotional processing are likely to impact on how pain is regulated.

6. SUMMARY

In summary, the recent data on the alterations in the TRYCAT pathway in somatization may allow for overlaps with data on the role of the mu-opioid, serotonin, and oxytocin systems in the regulation of early attachment. The effects of stress and the mu-opioid receptor on the TRYCAT pathway may contribute to the interaction between early attachment and somatization. Changes in the TRYCAT pathway, both centrally and peripherally, alter both pain and emotional processing, including via the regulation of the NMDAR and GPR35. The effects of somatization, and indeed early attachment, may be relevant modulators of events at later developmental time points. Such a conceptualization of somatization has treatment implications. Treatments for somatization are predominantly psychological, including psychotherapy and cognitive behavioral therapy (Maes et al., 2007), often coupled with antidepressant medications. The above would suggest a range of treatments for somatization that may be useful, especially in conjunction with antidepressants or psychotherapy. Theoretically promising avenues will be discussed in the succeeding section.

7. POTENTIAL TREATMENT IMPLICATIONS

Evidence suggests the usefulness of St. John's wort (*Hypericum perforatum*) in the treatment of somatization (Müller, Mannel, Murck, & Rahlfs, 2004). Interestingly, St. John's wort has higher levels of KA, 33.75 μg/tablet, than any other herbal medicine tested (Turski et al., 2011). High levels of KA were also found in dandelion leaves, also used as herbal medicine (Turski et al., 2011).

Resveratrol has a role in the inhibition of pain and hyperalgesia (Utreras, Terse, Keller, Iadarola, & Kulkarni, 2011). These effects seem to be mediated centrally (Falchi, Bertelli, Galazzo, Viganò, & Dib, 2010). Resveratrol inhibits the AHr, therefore inhibiting IDO induction (MacPherson & Matthews, 2010). In one study, resveratrol was shown to act as a competitive partial agonist of the AHr, inhibiting full agonist activation of AHr-induced genes (Bachleda, Vrzal, & Dvorak, 2010). If the increase in the KY/KA ratio and the activation of the AHr are indeed significant events in somatization, then this could suggest some efficacy of resveratrol in its treatment. Resveratrol may induce analgesia via the opioidergic system (Gupta, Sharma, & Briyal, 2004). As to whether this has relevance to the role of the mu-opioid receptor in specific CNS areas in the etiology of insecure attachment and somatization remains to be investigated. Resveratrol also has direct effects in monocytes, inhibiting LPS-induced IL-8, NF-KB, and COX2, suggesting an inhibitory effect on the putative monocyte role in somatization (Oh et al., 2009). The level of NF-KB induction in monocytes closely parallels increases in IDO, suggesting a role for resveratrol in the inhibition of monocyte IDO.

Melatonin is a chronobiotic and antioxidant. Experimental and clinical data support the analgesic role of melatonin, in a dose-dependent manner. Melatonin has analgesic benefits in patients with chronic pain, including fibromyalgia, irritable bowel syndrome, and migraine. The physiologic mechanism underlying the analgesic actions of melatonin has not been clarified. The effects may be linked to Gi-coupled melatonin receptors, to Gi-coupled mu-opioid receptors or GABA-B receptors with consequential decreases in anxiety and pain (Odagaki, Nishi, & Koyama, 2000). Mu-opioid receptors and melatonin membrane receptors, being Gi-protein coupled, decrease second messenger cAMP levels (Chneiweiss, Glowinski, & Premont, 1988; Nash & Osborne, 1995). GABA-B receptor agonists have been shown to have analgesic properties (McCarson & Enna, 1999; Patel et al., 2001). Melatonin may increase analgesia via the modulation of opioids and GABA-B receptors. Melatonin has also been shown to increase β-endorphin levels, further enhancing opioidergic activity (Shavali et al., 2005). The depletion of tryptophan by increased IDO and TDO will decrease levels of melatonin and serotonin. Melatonin improves sleep, reducing anxiety, which lowers pain levels (Wilhelmsen, Amirian, Reiter, Rosenberg, & Gogenur, 2011). Melatonin's antioxidant and anti-inflammatory effects in the periphery undoubtedly contribute to its analgesic efficacy. As to whether this efficacy includes the regulation of peripheral IDO and TDO requires investigation, as does any regulation of the GPR35. Melatonin increases the Th1 immune response, dampening the immunosuppression by IDO-induced regulatory T cells, and inhibits the effects of cortisol, likely via increases in bcl-2 associated anthanogene-1 (Quiros et al., 2008). As such, it seems likely that melatonin would generally inhibit TRYCAT pathway induction. However, in a study of rheumatoid arthritis, melatonin over 6 months increased the KY/KA ratio. Melatonin has undoubted antinociceptive benefits; however, its efficacy in somatization requires clarification.

Given that an increase in IDO and TDO drives tryptophan away from both serotonin and melatonin production, a combination of melatonin adjuvant to SSRIs may be useful in the treatment of somatization, perhaps especially if depression is also present. In a double-blind and placebo-controlled study in 101 patients to evaluate different doses of melatonin alone or in combination with fluoxetine for the management of fibromyalgia, it was found that both showed efficacy in improving pain, fatigue, rest/sleep, stiffness, and depression when administered alone (Hussain, Al-Khalifa, Jasim, & Gorial, 2011). A combination of melatonin and fluoxetine showed a significant reduction in anxiety and fatigue, and very significantly reduced depressive symptoms. As to whether adjuvant use of melatonin with fluoxetine would similarly improve somatization, including when comorbid with MDD, requires investigation.

Agomelatine is a melatonin receptor MT1r and MT2r agonist and serotonin 5HT-2Cr antagonist and is used as an antidepressant and anxiolytic (Owen, 2009). It is likely to have many similar effects to melatonin.

However, its activity at the 5HT-2Cr may also be relevant. Allelic variations in the 5HT-2Cr show a significant correlation with somatization in a patient sample (Ribasés et al., 2008). Alleles of the 5HT-2Cr also significantly modulate the efficacy of SSRIs in the treatment of neuropathic pain (Brasch-Andersen et al., 2011).

3-Hydroxykynurenine and QA toxicity are reversed by both MK801 and *N*-acetyl cysteine (NAC) (Nakagami, Saito, & Katsuki, 1996). The latter has potential as an antidepressant agent, particularly in bipolar disorder (Berk et al., 2008, 2011; Dean, Giorlando, & Berk, 2011; Magalhães et al., 2011). Tryptophan-derived metabolites induce T cell apoptosis, and this is inhibited by NAC (Lee et al., 2010). NAC has efficacy in inflammatory-based pain conditions (Perez et al., 2010). It also has antagonist effects on the NMDA system, blocking the effect of the NMDA system on the generation of reactive oxygen species (Tuneva, Bychkova, & Boldyrev, 2003). As such, it is a promising agent for somatization.

γ-Mangostin is a xanthone found in the fruit hulls of *Garcinia mangostana* L., which have long been used in Southeast Asia as a traditional medicine for the treatment of abdominal pain, dysentery, suppuration, wound infections, fever, chronic ulcer, and convulsions. Recent studies show that γ-mangostin exhibits a variety of pharmacological activities, including antagonism of the 5-HT-2A/C receptor, anti-inflammatory effects, and analgesic effects (Sukma, Tohda, Suksamran, & Tantisira, 2011). γ-Mangostin inhibits both central and peripheral nociception (Cui et al., 2010). Given the high levels of GPR35 receptors in the gastrointestinal tract, it would be interesting to test as to whether γ-mangostin mediates its analgesic effects via the GPR35, and as to whether it would have efficacy, both centrally and peripherally, in somatization.

Peripherally supplied KA does not readily cross the BBB. Recent data on a KA derivative KA amide show that it readily crosses the BBB, being neuroprotective in the absence of cognitive deficits (Gellert et al., 2012). Its efficacy in humans and in somatization remains to be investigated.

An as yet unidentified, glia-depressing factor (GDF) is differentially evident in human CSF in different medical conditions. This GDF inhibits KAT-I and KAT-II, preventing KA formation (Baran, Kepplinger, & Draxler, 2010). As to whether GDF is increased in somatization, or modulated by suggested somatization treatments, is unknown.

Pamoic acid salts are used to produce long-acting pharmaceutical formulations of FDA-approved drugs (Coleman et al., 1985), and so must also satisfy FDA safety criteria. The findings of Zhao et al. (2010) suggest that pamoate salts (pamoic acid) may contribute directly to the clinical effectiveness of some FDA-approved drugs through previously unrecognized GPR35-related mechanisms. Like KA and other GPR35 agonists, pamoic acid is likely to significantly modulate hyperalgesic responses. As to whether this would be relevant in different expressions of somatization remains to be examined.

The above treatment implications emphasize the importance of accurate diagnostic classification of somatization and its differentiation from depression, opening the door to a psychiatric classification based on physiological underpinnings (Maes & Rief, 2012).

REFERENCES

Anderson, G. (2011). Neuronal-immune interactions in mediating stress effects in the etiology and course of schizophrenia: Role of the amygdala in developmental co-ordination. *Medical Hypotheses*, *76*(1), 54–60.

Andreica-Sandica, B., Panaete, A., Pascanu, R., Sarban, C., & Andreica, V. (2011). The association between Helicobacter pylori chronic gastritis, psychological trauma and somatization disorder: A case report. *Journal of Gastrointestinal and Liver Diseases*, *29*(1), 301–303.

Bachleda, P., Vrzal, R., & Dvorak, Z. (2010). Resveratrol enhances NK cell cytotoxicity: Possible role for aryl hydrocarbon receptor. *Journal of Cellular Physiology*, *225*(2), 289–290.

Baran, H., Kepplinger, B., & Draxler, M. (2010). Endogenous kynurenine aminotransferases inhibitor is proposed to act as "Glia Depressing Factor" (GDF). *International Journal of Tryptophan Research*, *3*, 13–22.

Barth, M. C., Ahluwalia, N., Anderson, T. J., Hardy, G. J., Sinha, S., Alvarez-Cardona, J. A., et al. (2009). Kynurenic acid triggers firm arrest of leukocytes to vascular endothelium under flow conditions. *Journal of Biological Chemistry*, *284*, 19189–19195.

Berk, M., Copolov, D. L., Dean, O., Lu, K., Jeavons, S., Schapkaitz, I., et al. (2008). N-acetyl cysteine for depressive symptoms in bipolar disorder—A double-blind randomized placebo-controlled trial. *Biological Psychiatry*, *64*(6), 468–475.

Berk, M., Dean, O., Cotton, S. M., Gama, C. S., Kapczinski, F., Fernandes, B. S., et al. (2011). The efficacy of N-acetylcysteine as an adjunctive treatment in bipolar depression: An open label trial. *Journal of Affective Disorders*, *135*(1–3), 389–394.

Boess, F. G., De Vry, J., Erb, C., Flessner, T., Hendrix, M., Luithle, J., et al. (2007). The novel alpha7 nicotinic acetylcholine receptor agonist N-[(3R)-1-azabicyclo [2.2.2] oct-3-yl]-7-[2-(methoxy)phenyl]-1-benzofuran-2-carboxamide improves working and recognition memory in rodents. *The Journal of Pharmacology and Experimental Therapeutics*, *321*(2), 716–725.

Bonaccorso, S., Marino, V., Puzella, A., Pasquini, M., Biondi, M., Artini, M., et al. (2002). Increased depressive ratings in patients with hepatitis C receiving interferon-alpha-based immunotherapy are related to interferon-alpha-induced changes in the serotonergic system. *Journal of Clinical Psychopharmacology*, *22*(1), 86–90.

Booij, L., Swenne, C. A., Brosschot, J. F., Haffmans, P. M., Thayer, J. F., & Van der Does, A. J. (2006). Tryptophan depletion affects heart rate variability and impulsivity in remitted depressed patients with a history of suicidal ideation. *Biological Psychiatry*, *60*(5), 507–514.

Bowlby, J. (1969). Attachment and loss. In *Attachment* Vol. 1. The International Psycho-Analytic Library, 79:1-401. London: The Hogarth Press and the Institute of Psycho-Analysis.

Braidy, N., Grant, R., Adams, S., Brew, B. J., & Guillemin, G. J. (2009). Mechanism for quinolinic acid cytotoxicity in human astrocytes and neurons. *Neurotoxicity Research, 16*(1), 77–86.

Brasch-Andersen, C., Møller, M. U., Christiansen, L., Thinggaard, M., Otto, M., Brøsen, K., et al. (2011). A candidate gene study of serotonergic pathway genes and pain relief during treatment with escitalopram in patients with neuropathic pain shows significant association to serotonin receptor 2C (HTR2C). *European Journal of Clinical Pharmacology, 67*(11), 1131–1137.

Brown, F. W., Golding, J. M., & Smith, G. R.Jr., (1990). Psychiatric comorbidity in primary care somatization disorder. *Psychosomatic Medicine, 52*(4), 445–551.

Cairns, B. E., Svensson, P., Wang, K., Hupfeld, S., Graven-Nielsen, T., Sessle, B. J., et al. (2003). Activation of peripheral NMDA receptors contributes to human pain and rat afferent discharges evoked by injection of glutamate into the masseter muscle. *Journal of Neurophysiology, 90*(4), 2098–2105.

Chneiweiss, H., Glowinski, J., & Premont, J. (1988). Mu and delta opiate receptors coupled negatively to adenylate cyclase on embryonic neurons from the mouse striatum in primary cultures. *The Journal of Neuroscience, 8*, 3376–3382.

Chong, R. Y., Oswald, L., Yang, X., Uhart, M., Lin, P. I., & Wand, G. S. (2006). The mu-opioid receptor polymorphism A118G predicts cortisol responses to naloxone and stress. *Neuropsychopharmacology, 31*(1), 204–211.

Claes, S., Myint, A. M., Domschke, K., Del-Favero, J., Entrich, K., Endelberg, S., et al. (2011). The kynurenine pathway in major depression: Haplotype analysis of three related functional candidate genes. *Psychiatry Research, 188*(3), 355–360.

Coleman, M. D., Mihaly, G. W., Ward, S. A., Edwards, G., Howells, R. E., & Breckenridge, A. M. (1985). The sustained release of pyrimethamine base or pyrimethamine pamoate from a biodegradable injectable depot preparation in mice. *The Journal of Pharmacy and Pharmacology, 37*, 878–883.

Colombari, E., Sato, M. A., Cravo, S. L., Bergamaschi, C. T., Campos, R. R., Jr., & Lopes, O. U. (2001). Role of the medulla oblongata in hypertension. *Hypertension, 38*, 549–554.

Cosi, C., Mannaioni, G., Cozzi, A., Carlà, V., Sili, M., Cavone, L., et al. (2011). G-protein coupled receptor 35 (GPR35) activation and inflammatory pain: Studies on the antinociceptive effects of kynurenic acid and zaprinast. *Neuropharmacology, 60*, 1227–1231.

Cui, J., Hu, W., Cai, Z., Liu, Y., Li, S., Tao, W., et al. (2010). New medicinal properties of mangostins: Analgesic activity and pharmacological characterization of active ingredients from the fruit hull of Garcinia mangostana L. *Pharmacology, Biochemistry, and Behavior, 95*(2), 166–172.

Data-Franco, J., & Berk, M. (2012). The Nocebo effect: A clinician's guide. *The Lancet*, (in press).

Davies, S. J., Hood, S. D., Argyropoulos, S. V., Morris, K., Bell, C., Witchel, H. J., et al. (2006). Depleting serotonin enhances both cardiovascular and psychological stress reactivity in recovered patients with anxiety disorders. *Journal of Clinical Psychopharmacology, 26*(4), 414–418.

Dean, O., Giorlando, F., & Berk, M. (2011). N-acetylcysteine in psychiatry: Current therapeutic evidence and potential mechanisms of action. *Journal of Psychiatry & Neuroscience, 36*(2), 78–86.

Di Serio, C., Cozzi, A., Angeli, I., Doria, L., Micucci, I., Pellerito, S., et al. (2005). Kynurenic acid inhibits the release of the neurotrophic fibroblast growth factor (FGF)-1 and enhances proliferation of glia cells, in vitro. *Cellular and Molecular Neurobiology, 25*, 981–993.

Dignam, P., Parry, P., & Berk, M. (2010). Detached from attachment: Neurobiology and phenomenology have a human face. *Acta Neuropsychiatrica, 22*(4), 202–206.

Dougherty, D. M., Richard, D. M., James, L. M., & Mathias, C. W. (2010). Effects of acute tryptophan depletion on three different types of behavioral impulsivity. *International Journal of Tryptophan Research, 3*, 99–111.

Dozier, M., Stovall-McClough, C., & Albus, K. E. (2008). Attachment and psychopathology in adulthood. In J. Cassidy & P. R. Shaver (Eds.), *Handbook of attachment. Theory, research, and clinical implications* (pp. 718–744). (2nd ed.). New York and London: The Guilford Press.

Drummond, P. D. (2006). Tryptophan depletion increases nausea, headache and photophobia in migraine sufferers. *Cephalalgia, 26*(10), 1225–1233.

Euteneuer, F., Schwarz, M. J., Hennings, A., Riemer, S., Stapf, T., Selberdinger, V., et al. (2011). Psychobiological aspects of somatization syndromes: Contributions of inflammatory cytokines and neopterin. *Psychiatry Research, 195*(1–2), 60–65.

Falchi, M., Bertelli, A., Galazzo, R., Viganò, P., & Dib, B. (2010). Central antalgic activity of resveratrol. *Archives Italiennes de Biologie, 148*(4), 389–396.

Fink, P., Sorensen, L., Engberg, M., Holm, M., & Munk-Jorgensen, P. (1999). Somatization in primary care: Prevalence, health care utilization, and general practitioner recognition. *Psychosomatic Medicine, 40*(4), 330–338.

Funakoshi, H., Kanai, M., & Nakamura, T. (2011). Modulation of tryptophan metabolism, promotion of neurogenesis and alteration of anxiety-related behavior in tryptophan 2,3-dioxygenase-deficient mice. *International Journal of Tryptophan Research, 4*, 7–18.

Gellert, L., Varga, D., Ruszka, M., Toldi, J., Farkas, T., Szatmari, I., et al. (2012). Behavioural studies with a newly developed neuroprotective KYNA-amide. *Journal of Neural Transmission, 119*(2), 165–172.

Guillemin, G. J., Kerr, S. J., Smythe, G. A., Smith, D. G., Kapoor, V., Armati, P. J., et al. (2001). Kynurenine pathway metabolism in human astrocytes: A paradox for neuronal protection. *Journal of Neurochemistry, 78*(4), 842–853.

Guillemin, G. J., Smith, D. G., Kerr, S. J., Smythe, G. A., Kapoor, V., Armati, P. J., et al. (2000). Characterisation of kynurenine pathway metabolism in human astrocytes and implications in neuropathogenesis. *Redox Report, 5*, 108–111.

Guo, J., Williams, D. J., Puhl, H. L., III, & Ikeda, S. R. (2008). Inhibition of N-type calcium channels by activation of GPR35, an orphan receptor, heterologously expressed in rat sympathetic neurons. *The Journal of Pharmacology and Experimental Therapeutics, 324*, 342–351.

Gupta, Y. K., Sharma, M., & Briyal, S. (2004). Antinociceptive effect of trans-resveratrol in rats: Involvement of an opioidergic mechanism. *Methods and Findings in Experimental and Clinical Pharmacology, 26*(9), 667–672.

Han, Q., Cai, T., Tagle, D. A., & Li, J. (2010). Structure, expression, and function of kynurenine aminotransferases in human and rodent brains. *Cellular and Molecular Life Sciences, 67*(3), 353–368.

Han, Q., Robinson, H., Cai, T., Tagle, D. A., & Li, J. (2011). Biochemical and structural characterization of mouse mitochondrial aspartate aminotransferase, a newly identified kynurenine aminotransferase-IV. *Bioscience Reports, 31*(5), 323–332.

Hilmas, C., Pereira, E. F., Alkondon, M., Rassoulpour, A., Schwarcz, R., & Albuquerque, E. X. (2001). The brain metabolite kynurenic acid inhibits alpha7 nicotinic receptor activity and increases non-alpha7 nicotinic receptor expression: Physiopathological implications. *The Journal of Neuroscience, 21*(19), 7463–7473.

Huengsberg, M., Winer, J., Round, R., Gompels, M., & Shahmanesh, M. (1998). Elevated serum kynurenine level is associated with peripheral neuropathy in HIV. Neuroscience of HIV infection. *Journal of Neurovirology, 4*, 353.

Hussain, S. A., Al-Khalifa, I. I., Jasim, N. A., & Gorial, F. I. (2011). Adjuvant use of melatonin for treatment of fibromyalgia. *Journal of Pineal Research, 50*, 267–271.

Kaszaki, J., Palásthy, Z., Erczes, D., Rácz, A., Torday, C., Varga, G., et al. (2008). Kynurenic acid inhibits intestinal hypermotility and xanthine oxidase activity during experimental colon obstruction in dogs. *Neurogastroenterology and Motility*, *20*(1), 53–62.

Kilkens, T. O., Honig, A., van Nieuwenhoven, M. A., Riedel, W. J., & Brummer, R. J. (2004). Acute tryptophan depletion affects brain-gut responses in irritable bowel syndrome patients and controls. *Gut*, *53*(12), 1794–1800.

Klengel, T., Heck, A., Pfister, H., Bruckl, T., Hennings, J. M., Menke, A., et al. (2011). Somatization in major depression—Clinical features and genetic associations. *Acta Psychiatrica Scandinavica*, *124*(4), 317–328.

Koh, K. B., Kim, D. K., Kim, S. Y., & Park, J. K. (2005). The relation between anger expression, depression, and somatic symptoms in depressive disorders and somatoform disorders. *The Journal of Clinical Psychiatry*, *66*(4), 485–491.

Lapin, I. P. (1996). Anxiogenic activity of quinolinic acid and kynurenine in the social interaction test in mice. *Journal of Psychopharmacology*, *10*(3), 246–249.

Lapin, I. P. (1998). Antagonism of kynurenic acid to anxiogens in mice. *Life Sciences*, *63*(15), 231–236.

Laugeray, A., Launay, J. M., Callebert, J., Surget, A., Belzung, C., & Barone, P. R. (2010). Peripheral and cerebral metabolic abnormalities of the tryptophan-kynurenine pathway in a murine model of major depression. *Behavioural Brain Research*, *210*(1), 84–91.

Laugeray, A., Launay, J. M., Callebert, J., Surget, A., Belzung, C., & Barone, P. R. (2011). Evidence for a key role of the peripheral kynurenine pathway in the modulation of anxiety- and depression-like behaviours in mice: Focus on individual differences. *Pharmacology, Biochemistry, and Behavior*, *98*(1), 161–168.

Lee, S. M., Lee, Y. S., Choi, J. H., Park, S. G., Choi, I. W., Joo, Y. D., et al. (2010). Tryptophan metabolite 3-hydroxyanthranilic acid selectively induces activated T cell death via intracellular GSH depletion. *Immunology Letters*, *132*(1–2), 53–60.

Li, J., McRoberts, J. A., Ennes, H. S., Trevisani, M., Nicoletti, P., Mittal, Y., et al. (2006). Experimental colitis modulates the functional properties of NMDA receptors in dorsal root ganglia neurons. *American Journal of Physiology. Gastrointestinal and Liver Physiology*, *291*(2), G219–G228.

Liebau, C., Baltzer, A. W., Schmidt, S., Roesel, C., Karreman, C., Prisack, J. B., et al. (2002). Interleukin-12 and interleukin-18 induce indoleamine 2,3-dioxygenase (IDO) activity in human osteosarcoma cell lines independently from interferon-gamma. *Anticancer Research*, *22*(2A), 931–936.

Liu, L., Cohen, S., Schulz, M. S., & Waldinger, R. J. (2011). Sources of somatization: Exploring the roles of insecurity in relationships and styles of anger experience and expression. *Social Science and Medicine*, *73*, 1436–1443.

Luchowska, E., Kloc, R., Olajossy, B., Wnuk, S., Wielosz, M., Owe-Larsson, B., et al. (2009). Beta-adrenergic enhancement of brain kynurenic acid production mediated via cAMP-related protein kinase A signaling. *Progress in Neuropsychopharmacology and Biological Psychiatry*, *33*(3), 519–529.

MacPherson, L., & Matthews, J. (2010). Inhibition of aryl hydrocarbon receptor-dependent transcription by resveratrol or kaempferol is independent of estrogen receptor α expression in human breast cancer cells. *Cancer Letters*, *299*, 119–129.

Maes, M. (1995). Evidence for an immune response in major depression: A review and hypothesis. *Progress in Neuropsychopharmacology and Biological Psychiatry*, *19*(1), 11–38.

Maes, M. (2009). "Functional" or "psychosomatic" symptoms, e.g. a flu-like malaise, aches and pain and fatigue, are major features of major and in particular of melancholic depression. *Neuro Endocrinology Letters*, *30*(5), 564–573.

Maes, M., Galecki, P., Verkerk, R., & Rief, W. (2011). Somatization, but not depression, is characterized by disorders in the tryptophan catabolite (TRYCAT) pathway, indicating

increased indoleamine 2,3-dioxygenase and lowered kynurenine aminotransferase activity. *Neuroendocrinology Letters*, *32*(3), 101–111.
Maes, M., Leonard, B. E., Myint, A. M., Kubera, M., & Verkerk, R. (2011). The new 5-HT hypothesis of depression: Cell-mediated immune activation induces indoleamine 2,3-dioxygenase, which leads to lower plasma tryptophan and synthesis of detrimental tryptophan catabolites (TRYCATs), both of which contribute to the onset of depression. *Progress in Neuropsychopharmacology and Biological Psychiatry*, *35*(3), 702–721.
Maes, M., & Meltzer, H. Y. (1995). The serotonin hypothesis of major depression. Selected chapters on mood disorders. In F. Bloom & D. Kupfer (Eds.), *Psychopharmacology, the fourth generation of progress* (pp. 933–944). New York: Raven Press.
Maes, M., Mihaylova, I., Ruyter, M. D., Kubera, M., & Bosmans, E. (2007). The immune effects of TRYCATs (tryptophan catabolites along the IDO pathway): Relevance for depression—And other conditions characterized by tryptophan depletion induced by inflammation. *Neuroendocrinology Letters*, *28*(6), 826–831.
Maes, M., & Rief, W. (2012). Diagnostic classifications in depression and somatization should include biomarkers, such as disorders in the tryptophan catabolite (TRYCAT) pathway. *Psychiatric Research*, (Epub ahead of print).
Magalhães, P. V., Dean, O. M., Bush, A. I., Copolov, D. L., Malhi, G. S., Kohlmann, K., et al. (2011). N-acetyl cysteine add-on treatment for bipolar II disorder: A subgroup analysis of a randomized placebo-controlled trial. *Journal of Affective Disorders*, *129*(1–3), 317–320.
Martin, M. M., Liu, Y., & Wang, Z. (2012). Developmental exposure to a serotonin agonist produces subsequent behavioral and neurochemical changes in the adult male prairie vole. *Physiology and Behavior*, *105*(2), 529–535.
McCarson, K. E., & Enna, S. J. (1999). Nociceptive regulation of GABA(B) receptor gene expression in rat spinal cord. *Neuropharmacology*, *38*, 1767–1773.
McGinty, V. B., & Grace, A. A. (2009a). Timing-dependent regulation of evoked spiking in nucleus accumbens neurons by integration of limbic and prefrontal cortical inputs. *Journal of Neurophysiology*, *101*(4), 1823–1835.
McGinty, V. B., & Grace, A. A. (2009b). Activity-dependent depression of medial prefrontal cortex inputs to accumbens neurons by the basolateral amygdala. *Neuroscience*, *162*(4), 1429–1436.
McRoberts, J. A., Coutinho, S. V., Marvizón, J. C., Grady, E. F., Tognetto, M., Sengupta, J. N., et al. (2001). Role of peripheral N-methyl-D-aspartate (NMDA) receptors in visceral nociception in rats. *Gastroenterology*, *120*(7), 1737–1748.
Moir, A. T., & Eccleston, D. (1968). The effects of precursor loading in the cerebral metabolism of 5-hydroxyindoles. *Journal of Neurochemistry*, *15*(10), 1093–1108.
Mukai, M., Lin, T.-M., Peterson, R. E., Cooke, P. S., & Tischkau, S. A. (2008). Behavioral rhythmicity of mice lacking AhR and attenuation of light-induced phase shift by 2,3,7,8-tetracholordibenzo-p-dioxin. *Journal of Biological Rhythms*, *23*(3), 200–210.
Mukai, M., & Tischkau, S. A. (2007). Effects of tryptophan photoproducts in the circadian timing system: Searching for a physiological role for aryl hydrocarbon receptor. *Journal of Toxicological Science*, *95*(1), 172–181.
Müller, T., Mannel, M., Murck, H., & Rahlfs, V. W. (2004). Treatment of somatoform disorders with St. John's wort: A randomized, double-blind and placebo-controlled trial. *Psychosomatic Medicine*, *66*(4), 538–547.
Nakagami, Y., Saito, H., & Katsuki, H. (1996). 3-Hydroxykynurenine toxicity on the rat striatum in vivo. *Japanese Journal of Pharmacology*, *71*(2), 183–186.
Nash, M. S., & Osborne, N. N. (1995). Pertussis toxin-sensitive melatonin receptors negatively coupled to adenylate cyclase associated with cultured human and rat retinal pigment epithelial cells. *Investigative Ophthalmology and Visual Science*, *36*, 95–102.

Németh, H., Toldi, J., & Vécsei, L. (2005). Role of kynurenines in the central and peripheral nervous systems. *Current Neurovascular Research*, *2*(3), 249–260.

Niimi, S., Nakamura, T., Nawa, K., & Ichihara, A. (1983). Hormonal regulation of translatable RNA of tryptophan 2,3-dioxygenase in primary cultures of adult rat hepatocytes. *Journal of Biochemistry*, *94*, 1697–1706.

Noyes, R., Stuart, S., Langbehn, D. R., Happel, R. L., Longley, S. L., Muller, B. A., et al. (2003). Test of an interpersonal model of hypochondriasis. *Psychosomatic Medicine*, *65*, 292–300.

Odagaki, Y., Nishi, N., & Koyama, T. (2000). Functional coupling of GABA(B) receptors with G proteins that are sensitive to N-ethylmaleimide treatment, suramin, and benzalkonium chloride in rat cerebral cortical membranes. *Journal of Neural Transmission*, *107*, 1101–1116.

Oh, Y. C., Kang, O. H., Choi, J. G., Chae, H. S., Lee, Y. S., Brice, O. O., et al. (2009). Anti-inflammatory effect of resveratrol by inhibition of IL-8 production in LPS-induced THP-1 cells. *The American Journal of Chinese Medicine*, *37*(6), 1203–1214.

Ohira, K., Hagihara, H., Toyama, K., Takao, K., Kanai, M., Funakoshi, H., et al. (2010). Expression of tryptophan 2,3-dioxygenase in mature granule cells of the adult mouse dentate gyrus. *Molecular Brain*, *3*, 26.

Opitz, C. A., Litzenburger, U. M., Sahm, F., Ott, M., Tritschler, I., Trump, S., et al. (2011). An endogenous tumour-promoting ligand of the human aryl hydrocarbon receptor. *Nature*, *478*(7368), 197–203.

Orlikov, A. B., Prakh'e, I. B., & Ryzhov, I. V. (1990). Plasma kynurenine levels and the dexamethasone test in patients with endogenous anxiety disorders and depression. *Zhurnal Nevropatologii i Psikhiatrii Imeni S.S. Korsakova*, *90*(10), 83–85.

Orlikov, A., & Ryzov, I. (1991). Caffeine-induced anxiety and increase of kynurenine concentration in plasma of healthy subjects: A pilot study. *Biological Psychiatry*, *29*(4), 391–396.

Ortiz-Miranda, S., Dayanithi, G., Custer, E., Treistman, S. N., & Lemos, J. R. (2005). Micro-opioid receptor preferentially inhibits oxytocin release from neurohypophysial terminals by blocking R-type Ca2+ channels. *Journal of Neuroendocrinology*, *17*(9), 583–590.

Owen, R. T. (2009). Agomelatine: A novel pharmacological approach to treating depression. *Drugs of Today (Barcelona, Spain)*, *45*(8), 599–608.

Oxenkrug, G. F. (2007). Genetic and hormonal regulation of tryptophan kynurenine metabolism: Implications for vascular cognitive impairment, major depressive disorder, and aging. *Annals of the New York Academy of Sciences*, *1122*, 35–49.

Patel, S., Naeem, S., Kesingland, A., Froestl, W., Capogna, M., Urban, L., et al. (2001). The effects of GABA(B) agonists and gabapentin on mechanical hyperalgesia in models of neuropathic and inflammatory pain in the rat. *Pain*, *90*, 217–226.

Perez, R. S., Zollinger, P. E., Dijkstra, P. U., Thomassen-Hilgersom, I. L., Zuurmond, W. W., Rosenbrand, K. C., et al. (2010). Evidence based guidelines for complex regional pain syndrome type 1. *BMC Neurology*, *10*, 20.

Pezet, S., & McMahon, S. B. (2006). Neurotrophins: Mediators and modulators of pain. *Annual Review of Neuroscience*, *29*, 507–538.

Picciotto, M. R., Zoli, M., Léna, C., Bessis, A., Lallemand, Y., Le Novère, N., et al. (1995). Abnormal avoidance learning in mice lacking functional high-affinity nicotine receptor in the brain. *Nature*, *374*(6517), 65–67.

Prossin, A. R., Koch, A. E., Campbell, P. L., McInnis, M. G., Zalcman, S. S , & Zubieta, J. K. (2011). Association of plasma interleukin-18 levels with emotion regulation and μ-opioid neurotransmitter function in major depression and healthy volunteers. *Biological Psychiatry*, *69*(8), 808–812.

Quiros, I., Mayo, J. C., Garcia-Suarez, O., Hevia, D., Martin, V., Rodriguez, C., et al. (2008). Melatonin prevents the glucocorticoid receptor inhibition of cell proliferation and toxicity in hippocampal cells by reducing the glucocorticoid receptor nuclear translocation. *The Journal of Steroid Biochemistry and Molecular Biology, 110*, 116–124.

Ren, S., & Correia, M. A. (2000). Heme: A regulator of rat hepatic tryptophan 2,3-dioxygenase. *Archives of Biochemistry and Biophysics, 377*(1), 195–203.

Ribasés, M., Fernández-Aranda, F., Gratacòs, M., Mercader, J. M., Casasnovas, C., Núñez, A., et al. (2008). Contribution of the serotoninergic system to anxious and depressive traits that may be partially responsible for the phenotypical variability of bulimia nervosa. *Journal of Psychiatric Research, 42*(1), 50–57.

Rief, W., Pilger, F., Ihle, D., Bosmans, E., Egyed, B., & Maes, M. (2001). Immunological differences between patients with major depression and somatization syndrome. *Psychiatry Research, 105*(3), 165–174.

Rief, W., Shaw, R., & Fichter, M. (1998). Elevated levels of psychophysiological arousal and cortisol in patients with somatization syndrome. *Psychosomatic Medicine, 60*, 198–203.

Ruzicka, B. B., & Akil, H. (1997). The interleukin-1beta-mediated regulation of proenkephalin and opioid receptor messenger RNA in primary astrocyte-enriched cultures. *Neuroscience, 79*(2), 517–524.

Salm, A. K., Pavelko, M., Krouse, E. M., Webster, W., Kraszpulski, M., & Birkle, D. L. (2004). Lateral amygdaloid nucleus expansion in adult rats is associated with exposure to prenatal stress. *Brain Research. Developmental Brain Research, 148*(2), 159–167.

Sapko, M. T., Guidetti, P., Yu, P., Tagle, D. A., Pellicciari, R., & Schwarcz, R. (2006). Endogenous kynurenate controls the vulnerability of striatal neurons to quinolinate: Implications for Huntington's disease. *Experimental Neurology, 197*(1), 31–40.

Schmitt, M. L., Graeff, F. G., & Carobrez, A. P. (1990). Anxiolytic effect of kynurenic acid microinjected into the dorsal periaqueductal gray matter of rats placed in the elevated plus-maze test. *Brazilian Journal of Medical and Biological Research, 23*(8), 677–679.

Shavali, S., Ho, B., Govitrapong, P., Sawlom, S., Ajjimaporn, A., Klongpanichapak, S., et al. (2005). Melatonin exerts its analgesic actions not by binding to opioid receptor subtypes but by increasing the release of beta-endorphin an endogenous opioid. *Brain Research Bulletin, 64*(6), 471–479.

Shayit, M., Nowak, R., Keller, M., & Weller, A. (2003). Establishment of a preference by the newborn lamb for its mother: The role of opioids. *Behavioral Neuroscience, 117*(3), 446–454.

Silverman, W. R., de Rivero Vaccari, J. P., Locovei, S., Qiu, F., Carlsson, S. K., Scemes, E., et al. (2009). The pannexin 1 channel activates the inflammasome in neurons and astrocytes. *Journal of Biological Chemistry, 284*(27), 18143–18151.

Soria, V., Martínez-Amorós, E., Escaramís, G., Valero, J., Pérez-Egea, R., García, C., et al. (2010). Differential association of circadian genes with mood disorders: CRY1 and NPAS2 are associated with unipolar major depression and CLOCK and VIP with bipolar disorder. *Neuropsychopharmacology, 35*, 1279–1289.

Sroufe, L. A. (1995). *Emotional development. The organization of emotional life in the early years.* Cambridge: Cambridge University Press.

Steiner, J., Walter, M., Gos, T., Guillemin, G. J., Bernstein, H. G., Sarnyai, Z., et al. (2011). Severe depression is associated with increased microglial quinolinic acid in subregions of the anterior cingulate gyrus: Evidence for an immune-modulated glutamatergic neurotransmission? *Journal of Neuroinflammation, 8*, 94.

Stone, T. W., & Darlington, L. G. (2002). Endogenous kynurenines as targets for drug discovery and development. *Nature Reviews. Drug Discovery, 1*(8), 609–620.

Strathearn, L. (2011). Maternal neglect: Oxytocin, dopamine and the neurobiology of attachment. *Journal of Neuroendocrinology, 23*(11), 1054–1065.

Struzik, L., Duffin, J., Vermani, M., Hegadoren, K., & Katzman, M. A. (2002). Effects of tryptophan depletion on central and peripheral chemoreflexes in man. *Respiratory Physiology & Neurobiology*, *133*(3), 183–195.

Sukma, M., Tohda, M., Suksamran, S., & Tantisira, B. (2011). γ-Mangostin increases serotonin 2A/2C, muscarinic, histamine and bradykinin receptor mRNA expression. *Journal of Ethnopharmacology*, *135*(2), 450–454.

Surcinelli, P., Rossi, N., Montebarocci, O., & Baldaro, B. (2010). Adult attachment styles and psychological disease: Examining the mediating role of personality traits. *The Journal of Psychology*, *144*(6), 523–534.

Swartz, K. J., During, M. J., Freese, A., & Beal, M. F. (1990). Cerebral synthesis and release of kynurenic acid: An endogenous antagonist of excitatory amino acid receptors. *The Journal of Neuroscience*, *10*(9), 2965–2973.

Talbot, J. N., Roman, D. L., Clark, M. J., Roof, R. A., Tesmer, J. J. G., Neubig, R. R., et al. (2010). Differential modulation of mu-opioid receptor signaling to adenylyl cyclase by RGS proteins 4 or 8 and 7 in permeabilised C6 cells is Gα subtype dependent. *Journal of Neurochemistry*, *112*(4), 1026–1034.

Ting, K. K., Brew, B. J., & Guillemin, G. J. (2009). Effect of quinolinic acid on human astrocytes morphology and functions: Implications in Alzheimer's disease. *Journal of Neuroinflammation*, *6*, 36.

Ting, P., Cushenberry, P. A., Friedman, T. C., & Loh, Y. P. (1997). Enhanced brain opioid receptor activity precedes blood–brain barrier disruption. *Acta Neurochirurgica. Supplement*, *70*, 250–253.

Tuneva, E. O., Bychkova, O. N., & Boldyrev, A. A. (2003). Effect of NMDA on production of reactive oxygen species by human lymphocytes. *Bulletin of Experimental Biology and Medicine*, *136*, 159–161.

Turski, M. P., Turska, M., Zgrajka, W., Bartnik, M., Kocki, T., & Turski, W. A. (2011). Distribution, synthesis, and absorption of kynurenic acid in plants. *Planta Medica*, *77*(8), 858–864.

Unschuld, P. G., Ising, M., Roeske, D., Erhardt, A., Specht, M., Kloiber, S., et al. (2010). Gender-specific association of galanin polymorphisms with HPA-axis dysregulation, symptom severity, and antidepressant treatment response. *Neuropsychopharmacology*, *35*(7), 1583–1592.

Utreras, E., Terse, A., Keller, J., Iadarola, M. J., & Kulkarni, A. B. (2011). Resveratrol inhibits Cdk5 activity through regulation of p35 expression. *Molecular Pain*, *7*, 49.

van Veen, J. F., van Vliet, I. M., de Rijk, R. H., van Pelt, J., Mertens, B., Fekkes, D., et al. (2009). Tryptophan depletion affects the autonomic stress response in generalized social anxiety disorder. *Psychoneuroendocrinology*, *34*(10), 1590–1594.

Vécsei, L., & Beal, M. F. (1990). Influence of kynurenine treatment on open-field activity, elevated plus-maze, avoidance behaviors and seizures in rats. *Pharmacology, Biochemistry, and Behavior*, *37*(1), 71–76.

Wang, J., Simonavicius, N., Wu, X., Swaminath, G., Reagan, J., Tian, H., et al. (2006). Kynurenic acid as a ligand for orphan G protein-coupled receptor GPR35. *Journal of Biological Chemistry*, *281*(31), 22021–22028.

Warnick, J. E., McCurdy, C. R., & Sufka, K. J. (2005). Opioid receptor function in social attachment in young domestic fowl. *Behavioural Brain Research*, *160*(2), 277–285.

Wearden, A. J., Lamberton, N., Crook, N., & Walsh, V. (2005). Adult attachment, alexithymia, and symptom reporting. An extension to the four category model of attachment. *Journal of Psychosomatic Research*, *58*(3), 279–288.

Wilhelmsen, M., Amirian, I., Reiter, R. J., Rosenberg, J., & Gogenur, I. (2011). Analgesic effects of melatonin: A review of current evidence from experimental and clinical studies. *Journal of Pineal Research*, *51*, 270–277.

Wu, L. J., & Zhuo, M. (2009). Targeting the NMDA receptor subunit NR2B for the treatment of neuropathic pain. *Neurotherapeutics*, *6*(4), 693–702.

Yang, X., Yang, H. B., Xie, Q. J., Liu, X. H., & Hu, X. D. (2009). Peripheral inflammation increased the synaptic expression of NMDA receptors in spinal dorsal horn. *Pain*, *144*(1–2), 162–169.

Zepf, F. D., Holtmann, M., Stadler, C., Wöckel, L., & Poustka, F. (2009). Reduced serotonergic functioning changes heart rate in ADHD. *Journal of Neural Transmission*, *116*(1), 105–108.

Zhang, J., Wu, Z., Zhou, L., Li, H., Teng, H., Dai, W., et al. (2011). Deficiency of antinociception and excessive grooming induced by acute immobilization stress in Per1 mutant mice. *PLoS One*, *6*(1), e16212.

Zhao, P., Sharir, H., Kapur, A., Cowan, A., Geller, E. B., Adler, M. W., et al. (2010). Targeting of the orphan receptor GPR35 by pamoic acid: A potent activator of extracellular signal-regulated kinase and beta-arrestin2 with antinociceptive activity. *Molecular Pharmacology*, *78*, 560–568.

CHAPTER THREE

Inflammation in Schizophrenia

Norbert Müller[1], Aye-Mu Myint, Markus J. Schwarz
Department of Psychiatry and Psychotherapy, Ludwig-Maximilians-Universität München, München, Germany
[1]Corresponding author: e-mail address: Norbert.Mueller@med.uni-muenchen.de

Contents

Abstract

Although there is no doubt that the dopaminergic neurotransmission is strongly involved in the pathophysiology of schizophrenia, the exact mechanism leading to dopaminergic dysfunction is still unclear. A disbalance in the immune response associated with a slight inflammatory process of the central nervous system (CNS) has been postulated. Such a mechanism is the basis for the "mild encephalitis" concept. A dysfunction in the activation of the type-1 immune response seems to be associated with decreased activity of the key enzyme of the tryptophan/kynurenine metabolism, indoleamine 2,3-dioxygenase (IDO). Theoretically, a decreased activity of IDO results in the increased production of kynurenic acid, an *N*-methyl-D-aspartate antagonist in the CNS, and a reduced glutamatergic neurotransmission in schizophrenia. Accordingly, in animal models of schizophrenia, increased levels of kynurenic acid in critical regions of the CNS were described, although studies of peripheral blood levels of kynurenic acid in schizophrenic patients showed controversial results. The immunological effects of a lot of existing antipsychotics, however, rebalance in part the immune imbalance and the overweight of the production of kynurenic acid. The inflammatory state in schizophrenia is associated with increased prostaglandin E_2 production and increased cyclooxygenase-2 (COX-2) expression. Growing evidence from clinical studies with

Advances in Protein Chemistry and Structural Biology, Volume 88
ISSN 1876-1623
http://dx.doi.org/10.1016/B978-0-12-398314-5.00003-9

COX-2 inhibitors points to favorable effects of anti-inflammatory therapy in schizophrenia, in particular in an early stage of the disorder. Further options for immunomodulating therapies in schizophrenia will be discussed.

1. INTRODUCTION

Pathophysiological studies in schizophrenia were focused on disturbances of the dopaminergic neurotransmission over the past 50 years without convincing results. Although there is no doubt that a disturbance of the dopaminergic neurotransmission is involved in the pathophysiology of schizophrenia, a further elucidation of the mechanisms of this disturbance showed unsatisfactory results over the past 20–30 years. Antidopaminergic drugs, the existing typical and atypical antipsychotics, are effective in schizophrenia, namely in the therapy of schizophrenic positive symptoms. Nevertheless, the effects on schizophrenic negative symptoms, on cognitive deficits, and other deficit symptoms show disappointing therapeutic effects. The chronic course of at least one-third of schizophrenia patients, described before the era of antipsychotics, seems not improved in the majority of chronic patients despite progress in social-psychiatric reintegration programs, cognitive training, psychoeducational approaches, and other rehabilitation concepts.

New concepts in the biological research of schizophrenia for a better approach to the pathophysiological mechanisms are required (Müller & Schwarz, 2012). Recently, genetic data from multiple large cohorts of patients showed that different gene loci located on chromosome 6p22.1 are the most probable susceptibility genes for schizophrenia (Purcell et al., 2009; Shi et al., 2009; Stefansson et al., 2009). The region includes several genes of interest, which are related to the immune function. The strongest evidence for association was observed in or near a cluster of histone protein genes which could be relevant through their roles in regulation of DNA transcription or repair, that is, in epigenetics (Costa et al., 2007), or their direct role in antimicrobial defense (Kawasaki & Iwamuro, 2008). Moreover, several genes of the HLA complex, which regulate the immune function and already earlier have been discussed to be involved in schizophrenia, are located in these regions (Fellerhoff, Laumbacher, Mueller, Gu, & Wank, 2007).

Although an immune dysfunction and the involvement of infectious agents in the pathophysiology of schizophrenia are discussed since decades, the field never came into the mainstream of research. These genetic findings and further recent interesting observations, however, may contribute shifting research into the direction of immunological alterations and inflammation as cause for schizophrenia.

Infectious agents such as cytomegalo-, influenza-, borna-virus, Chlamydia, *Toxoplasma gondii*, and many others have been discussed to be involved in schizophrenia, but results from animal models of schizophrenia indicate that not a certain virus or other infectious agent but the immune response determines the risk for schizophrenia: Immune stimulants such as lipopolysaccharides or poly I:C, which mimic a bacterial or viral infection both lead to typical behavior alterations in the offsprings of animals (Winter et al., 2009; Zuckerman & Weiner, 2005). In humans, it could be shown that increased maternal levels of the proinflammatory cytokine interleukin-8 (IL-8) during pregnancy are associated with an increased risk for schizophrenia in the offspring—whatever the reason for increased IL-8 was (Brown, 2006).

There is no doubt that the dopaminergic neurotransmission plays an important role in schizophrenia. Dopaminergic hyperfunction in the limbic system and dopaminergic hypofunction in the frontal cortex are discussed to be the main neurotransmitter disturbances. Recent research provides further insight that glutamatergic hypofunction might be the cause for this dopaminergic dysfunction in schizophrenia (Swerdlow, van Bergeijk, Bergsma, Weber, & Talledo, 2009). The function of the glutamatergic system is closely related to the immune system and to the tryptophan–kynurenine metabolism, which both seem to play a key role in the pathophysiology of schizophrenia (Müller & Schwarz, 2007a, 2007b).

2. THE IMMUNE RESPONSE AND THE TYPE-1 AND TYPE-2 POLARIZATION

The innate immune system is the phylogenetic oldest part of the immune response, for example, natural killer cells and monocytes as the first barrier of the immune system being part of it. The adaptive immune response with the antibody producing B-lymphocytes, the T-lymphocytes, and their regulating "immunotransmitters," the cytokines, is the specifically acting component of the immune system. Cytokines regulate all types and all cellular components of the immune system including the innate immune system. Helper T-cells are of two types: T-helper-1 (TH-1) and T-helper-2 (TH-2). TH-1 cells produce the characteristic "type-1" activating cytokines such as interleukin-2 (IL-2) and interferon-γ (IFN-γ). However, since not only TH-1 cells but also certain monocytes/macrophages (M1) and other cell types produce these cytokines, the immune response is named type-1 immune response. The humoral—antibody producing—arm of the adaptive immune system is mainly activated by the type-2 immune response. TH-2 or certain monocytes/macrophages (M2) produce mainly IL-4, IL-10, and IL-13 (Mills, Kincaid, Alt, Heilman, & Hill, 2000). Another terminology separates the cytokines into proinflammatory and anti-inflammatory ones. Proinflammatory cytokines such as tumor necrosis factor-α (TNF-α) and IL-6 are primarily secreted from monocytes and macrophages, activating other cellular components of the inflammatory response. While TNF-α is a ubiquitous expressed cytokine mainly activating the type-1 response, IL-6 activates the type-2 response including the antibody production. Anti-inflammatory cytokines such as IL-4 and IL-10 help to downregulate the inflammatory immune response.

The type-1 immune system promotes the cell-mediated immune response directed against intracellular pathogens, whereas the type-2 response helps B-cell maturation and promotes the humoral immune response including the production of antibodies directed against extracellular pathogens. Type-1 and type-2 cytokines antagonize each other in promoting their own type of response, while suppressing the immune response of the other, therefore, the term "polarized" is used.

3. INFLAMMATION IN SCHIZOPHRENIA

Infection during pregnancy in mothers of offsprings later developing schizophrenia has been repeatedly described, in particular in the second trimester (Brown et al., 2004; Buka, Goldstein, Seidman, & Tsuang, 2000). As opposed to any single pathogen, the immune response itself of the mother may be related to the increased risk for schizophrenia in the offspring (Zuckerman & Weiner, 2005). In animal models, an immune activation of the mother during the second trimenon of pregnancy led to schizophrenia-like symptoms in the offspring in adulthood in a series of studies (Meyer, Schwarz, & Müller, 2011). In humans, increased IL-8 levels of mothers during the second trimenon were associated with an increased risk for schizophrenia in the offspring (Brown et al., 2004). A fivefold increased risk for developing psychoses later on, however, was detected after infection of the central nervous system (CNS) in early childhood (Brown et al., 2004; Gattaz, Abrahao, & Foccacia, 2004).

These data were confirmed in recent studies (Brown, 2008; Dalman et al., 2008; Koponen et al., 2004).

Signs of inflammation were found in schizophrenic brains (Körschenhausen, Hampel, Ackenheil, Penning, & Müller, 1996), and the term "mild localized chronic encephalitis" to describe a slight but chronic inflammatory process in schizophrenia was proposed (Bechter, 2001).

4. TYPE-1 AND TYPE-2 IMMUNE RESPONSE IN SCHIZOPHRENIA

Decreased levels of neopterin, a product of activated monocytes/macrophages, also point to a blunted activation of the type-1 response (Sperner-Unterweger, Miller, et al., 1999). The decreased response of lymphocytes after stimulation with specific antigens reflects a reduced capacity for a type-1 immune response in schizophrenia as well (Müller, Ackenheil, Hofschuster, Mempel, & Eckstein, 1991). Inter-cellular adhesion molecule-1 (ICAM-1) is a type-1-related protein, and a cell-adhesion molecule expressed on macrophages and lymphocytes. Decreased levels of the soluble ICAM-1 (sICAM-1), as found in schizophrenia, also represent an underactivation of the type-1 immune system (Schwarz, Riedel, Ackenheil, & Müller, 2000). Decreased levels of the soluble TNF-receptor p55—mostly decreased when TNF-α is decreased—were observed too (Haack et al., 1999). A blunted response of the skin to different antigens in schizophrenia was observed before the era of antipsychotics (Molholm, 1942). This finding could be replicated in unmedicated schizophrenic patients using a skin test of the cellular immune response (Riedel et al., 2007). However, there are some conflicting results indicating possibly increased levels of Th1 cytokines in schizophrenia (Bresee & Rapaport, 2009). The latest meta-analysis showed dominant proinflammatory changes in schizophrenia but not in the Th2 cytokines (Potvin et al., 2008). After including antipsychotic medication effects into the analysis—these effects on immune parameters are discussed below—only increases of IL-1 receptor antagonist serum levels and IL-6 serum levels were found. Type-1 parameters, hypothesized to be down-regulated in schizophrenia, were not included into the meta-analysis because only a few studies have been performed in unmedicated patients.

Several reports described increased serum IL-6 levels in schizophrenia (Cazzullo et al., 1998). IL-6 serum levels might be especially high in patients with an unfavorable course of the disease (Lin et al., 1998). IL-6 is a product of activated monocytes, and some authors refer it as a marker of the type-2 immune response. Moreover, several other signs of activation of the type-2 immune response are described in schizophrenia, including the increased Th2 type of lymphocytes in the blood (Sperner-Unterweger, Whitworth, et al., 1999), the increased production of immunoglobulin E, and an increase of IL-10 serum levels (Schwarz, Chiang, Müller, & Ackenheil, 2001; van Kammen, McAllister-Sistilli, & Kelley, 1997). In the cerebrospinal fluid (CSF), IL-10 levels were found to be related to the severity of the psychosis (van Kammen et al., 1997).

The key cytokine of the type-2 immune response is IL-4. Increased levels of IL-4 in the CSF of juvenile schizophrenic patients have been reported (Mittleman et al., 1997), which indicates that the increased type-2 response in schizophrenia is not only a phenomenon of the peripheral immune response.

The data show, however, that the immune response in schizophrenia is confounded by factors partly disease inherent such as duration of disease, chronicity, or therapy response and partly other factors such as antipsychotic medication and smoking.

5. ANTIPSYCHOTIC DRUGS AND THE TYPE-1/TYPE-2 IMBALANCE IN SCHIZOPHRENIA

In vitro studies show that the blunted IFN-γ production becomes normalized after therapy with antipsychotics (Wilke et al., 1996). An increase of "memory cells" ($CD4^+CD45RO^+$ cells)—one of the main sources of IFN-γ production—during antipsychotic therapy with neuroleptics was observed by different groups (Müller, Riedel, Schwarz, Gruber, & Ackenheil, 1997). Additionally, an increase of sIL-2R—the increase reflects an increase of activated IL-2-bearing T-cells—during antipsychotic treatment was described (Müller, Empl, Riedel, Schwarz, & Ackenheil, 1997). The reduced sICAM-1 levels show a significant increase during short-term antipsychotic therapy (Schwarz et al., 2000), and the ICAM-1 ligand leukocyte function antigen-1 shows a significantly increased expression during antipsychotic therapy (Müller et al., 1999). The increase of TNF-α and TNF-α receptors during therapy with clozapine was observed repeatedly (Pollmächer, Schuld, Kraus, Haack, & Hinze-Selch, 2001). Moreover, the blunted reaction to vaccination with *Salmonella typhi* was not observed in patients medicated with antipsychotics (Ozek, Toreci, Akkok, & Guvener, 1971). An elevation of IL-18 serum levels was described in medicated schizophrenics (Tanaka, Shintani, Fujii, Yagi, & Asai, 2000). Since IL-18 plays a pivotal role in the type-1 immune response, this finding is consistent with other descriptions of type-1 activation during antipsychotic treatment.

Regarding the type-2 response, several studies point out that antipsychotic therapy is accompanied by a functional decrease of the IL-6 system (Maes et al., 1997; Müller et al., 2000). These findings provide further evidence that antipsychotics have a "balancing" effect on cytokines.

6. THE MONOCYTE/MACROPHAGE SYSTEM IN SCHIZOPHRENIA

Monocytes are the most important part of the innate immune system and play a pivotal role as the first barrier against invading pathogens. When activated, monocytes turn into macrophages and invade all sorts of tissue, phagocytosing antigenic material and presenting antigens to the immune system. Monocytic cytokines such as IL-1, IL-6, and TNF-α activate the further cascade of the immune reaction, including the adaptive immune system of T- and B-lymphocytes. Macrophages that invade the CNS convert into microglia (Hanisch & Kettenmann, 2007). Microglia are the main component of the CNS immune system, make up about 12% of the CNS cells and have a high variability in morphology and function (Bernstein, Steiner, & Bogerts, 2009; Block, Zecca, & Hong, 2007). Toll-like receptors (TLRs) are expressed on human monocytes and microglia (Jack et al., 2005), involved in the regulation of the CNS immune response and suggested to be the link between the immune system and the CNS (Downes & Crack, 2010). Therefore, the specific role of TLRs in a disorder with a possible immunological or inflammatory component is of high interest.

Functional studies of monocytes have not been performed in schizophrenia; although earlier findings point to an alteration in numbers of monocytes, several studies have shown that unmedicated schizophrenia patients have a significantly higher number of monocytes than healthy controls (Bonartsev, 2008; Dameshek, 1930; N. Müller et al., unpublished observations; Rothermundt, Arolt, Weitzsch, Eckhoff, & Kirchner, 1998; Wilke et al., 1996), an observation that has also been made in juvenile schizophrenia (T. Falcone, personal communication). Regarding monocyte/macrophage function, levels of the macrophage activation marker neopterin were found to be lower in unmedicated schizophrenia patients than in controls (Sperner-Unterweger, Miller, et al., 1999), although findings are inconsistent (Chittiprol et al., 2010). Monocytic cytokines (IL-6, TNF-α) and neopterin are influenced by antipsychotic treatment (Korte et al., 1998;

Na & Kim, 2007; Sperner-Unterweger, Miller, et al., 1999) so that medication has to be taken into consideration in schizophrenia studies.

Interestingly, a specific monocytic inflammatory profile has also been described in schizophrenia, that is, an increased expression of inflammatory and chemotactic monocytic genes (including IL-6, IL-1β, TNF-α, and COX-2) (Drexhage et al., 2010). Moreover, in schizophrenia, alterations in monocyte-derived microglia cells have been observed (Rothermundt et al., 2004; Steiner et al., 2008). Lower stimulation of TLR-3 and TLR-4 in schizophrenia point to a functional disturbance of the monocytic system as an underlying immunological mechanism in schizophrenia (Müller, Wagner, et al., 2012). There are also descriptions of an increase in activated microglial cells (Bayer, Buslei, Havas, & Falkai, 1999; Radewicz, Garey, Gentleman, & Reynolds, 2000; Wierzba-Bobrowicz, Lewandowska, Lechowicz, Stepien, & Pasennik, 2005) as well as an accumulation of macrophages in the CSF of schizophrenic patients during an acute psychotic episode (Nikkilä, Muller, Ahokas, Rimon, & Andersson, 2001).

7. BRAIN IMAGING AND MICROGLIA ACTIVATION IN SCHIZOPHRENIA

Neuroinflammation is characterized by the activation of microglia cells, which show an increase in the expression of the peripheral benzodiazepine receptor. The isoquinoline (R)-N-(11)C-methyl-N-(1-methylpropyl)-1-(2-chlorophenyl)isoquinoline-3-carboxamide ((11)C-(R)-PK11195) is a peripheral benzodiazepine receptor ligand that can be used for the imaging of activated microglia cells, and thus neuroinflammation, with PET. In a study using the ligand (PK11195) for microglial activation, an increased expression of PK11195 as hint for an inflammatory process was shown in schizophrenia (van Berckel et al., 2008). In a further study, a significantly higher binding potential of (11)C-(R)-PK11195 was found in the hippocampus of schizophrenic patients than in healthy volunteers. A nonsignificant 30% higher (11)C-(R)-PK11195 binding potential was found in the whole-brain gray matter of schizophrenic patients (Doorduin et al., 2009). Increased expression of PK11195 was found, for example, in certain forms of encephalitis. This *in vivo* result from an imaging study supports further the view that an inflammatory process plays a role in schizophrenia. A study in 14 chronic schizophrenic patients using the PET-ligand (11)C-DAA1106 showed associations between microglia activation and positive symptom scores of the PANS-scale and the duration of the disease. That is, in chronic schizophrenic patients, microglia activation seems to be more pronounced (Takano et al., 2010). The role of antipsychotic medication in the activation of microglia, however, is discussed controversially.

8. THE TRYPTOPHAN–KYNURENINE METABOLISM IN SCHIZOPHRENIA

In contrast to microglial cells which produce quinolinic acidolinic acid, astrocytes play a key role in the production of KYNA in the CNS. Astrocytes are the main source of KYNA (Heyes, Chen, Major, & Saito, 1997). The cellular localization of the kynurenine metabolism is primarily in macrophages and microglial cells, but also in astrocytes (Kiss et al., 2003). KMO, however, a critical enzyme in the kynurenine metabolism, is absent in human astrocytes (Guillemin et al., 2001). Accordingly, it has been described that astrocytes cannot produce the product 3-hydroxykynurenine (3-HK), but they are able to produce large amounts of early kynurenine metabolites, such as KYN and KYNA (Guillemin et al., 2001). This supports the observation that inhibition of KMO leads to an increase of the KYNA production in the CNS (Chiarugi, Carpenedo, & Moroni, 1996). The complete metabolism of kynurenine to quinolinic acidolinic acid is observed mainly in microglial cells, only a small amount of quinolinic acidolinic acid is produced in astrocytes via a side arm of the kynurenine metabolism. Therefore, due to the lack of kynurenine-hydroxylase (KYN-OHse), in case of high tryptophan breakdown to KYN, KYNA may accumulate in astrocytes.

A second key player in the metabolization of 3-HK are monocytic cells infiltrating the CNS. They help astrocytes in the further metabolism to quinolinic acidolinic acid (Guillemin et al., 2001). However, the low levels of sICAM-1 (ICAM-1 is the molecule that mainly mediates the penetration of monocytes and lymphocytes into the CNS) in the serum and in the CSF of nonmedicated schizophrenic patients (Schwarz et al., 2000), and the increase of adhesion molecules during antipsychotic therapy indicate that the penetration of monocytes may be reduced in nonmedicated schizophrenic patients (Müller et al., 1999).

9. TREATMENT OPTIONS FOR SCHIZOPHRENIA BASED ON IMMUNE MODULATION

An immune-based therapeutic approach has been proposed decades ago: the Nobel-laureate Julius Ritter Wagner von Jauregg developed a vaccination therapy for psychoses (in those days not yet termed schizophrenia)

(Wagner von Jauregg, 1926). He treated patients successfully with vaccines of tuberculosis, malaria, or *S. typhi* stimulating the type-1 immune response (Müller, Schwarz, & Riedel, 2005). The immune-based "vaccination" therapy—although promising—was developed during the first decades of the twentieth century but not followed-up outside German-speaking countries, in particular after the introduction of electroconvulsive therapy and later the therapy with neuroleptics.

Beside anti-inflammatory drugs—in particular, the COX-2 inhibitors discussed beyond—very preliminary data exist for other therapeutics related to the immune function.

Several studies have been performed using omega-3 fatty acids in schizophrenia: the results are inconsistent until now, the overall effect size is small (Ross, Seguin, & Sieswerda, 2007) in comparison with placebo in first episode schizophrenia and in chronic schizophrenia. More intriguing is the result of the study of Amminger et al. (2010). They found in a 12-month study in a group of persons with a high risk for schizophrenia showing prodromal symptoms, a significantly lower transition rate to psychosis in persons who got the omega-3 fatty acid capsules compared to placebo-treated controls. The results of the studies in schizophrenia can be seen at Table 3.1.

Erythropoietin is a substance that shows immune-modulating effects, although other effects also play a role. A 12-week placebo-controlled treatment study with rh-erythropoietin in chronic schizophrenic patients was associated with a significant improved cognitive performance compared to placebo. No better outcome compared to placebo was found in the PANS overall psychopathology, nor in the positive and negative symptom subscales. Regarding the social functioning, no advance of erythropoietin could be observed either (Ehrenreich et al., 2007). Interestingly, rh-erythropoietin seems to delay the loss of CNS volume in schizophrenic patients (Wüstenberg et al., 2011).

Furthermore, there is recent evidence indicating that a tetracycline antibiotic is effective as an add-on treatment in schizophrenia (Chaves et al., 2009). Although minocycline has a wide range of actions in the brain, but most predominantly, it has anti-inflammatory action via modulating the nitric oxide system. Moreover, minocycline inhibits microglia activation. Microglia, the monocyte/macrophage-derived CNS cells have been shown to be more activated in schizophrenia than in healthy control (van Berckel et al., 2008). Case reports confirm the view of clinical effects of minocycline in schizophrenia (Ahuja & Carroll, 2007). In animal experiments, an effect of minocycline on cognitive functions has been observed (Mizoguchi et al., 2008).

Table 3.1 Double-blind, placebo-controlled study of omega-3 fatty acids in schizophrenia

Authors	Study design	Primary analysis	Secondary analysis
High risk			
Amminger et al. (2010)	Mono; EPA-rich oil versus placebo	EPA > placebo	
First episode			
Peet, Brind, Ramchand, Shah, and Vankar (2001)	Mono; 2 g: EPA versus placebo	EPA > placebo for antipsychotic drug requirement	
Berger (2004)	Add-on; 2 g: EPA versus placebo	EPA = placebo	EPA > placebo for antipsychotic dose
Chronic schizophrenia			
Peet et al. (2001)	Add-on; 2 g: EPA versus DHA versus placebo	EPA > placebo	EPA > DHA
Fenton, Dickerson, Boronow, Hibbeln, and Knable (2001)	Add-on; 3 g. EPA versus placebo	EPA = placebo	
Peet and Horrobin (2002)	Add-on; 1, 2, & 5 g: EPA versus placebo	EPA = placebo	EPA (2 g) > placebo in clozapine subgroup
Emsley, Myburgh, Oosthuizen, and van Rensburg (2002)	Add-on; 3 g: EA versus placebo	EPA > placebo	
Bentsen (2006)	Add-on; 2 g: EPA versus antioxidants versus combination versus placebo	EPA < placebo	EPA + antioxidants = placebo
Emsley et al. (2006)	Add-on, 2 g: EPA versus placebo	N/A (primary analysis for effect on tardive dyskinesia)	EPA = placebo

(adapted from Peet, 2008)

No benefit regarding the clinical symptoms of schizophrenia was found in a small group of schizophrenic patients which were tested seropositive for cytomegalo-virus positive in a double-blind study using the virustatic valacyclovir in a double-blind, randomized, prospective add-on design (Dickerson, Boronow, Stallings, Origoni, & Yolken, 2003; Dickerson, Stallings, Boronow, Origoni, Sullens, et al., 2009). Neither superiority compared to placebo was observed in schizophrenic patients showing seropositivity for toxoplasmosis during a trial of azithromycin (Dickerson, Stallings, Boronow, Origoni, & Yolken, 2009).

10. COX-2 INHIBITION AS THERAPEUTIC APPROACH IN SCHIZOPHRENIA

COX inhibition provokes differential effects on the kynurenine metabolism: while COX-1 inhibition increases the levels of KYNA, COX-2 inhibition decreases them (Schwieler, Erhardt, Erhardt, & Engberg, 2005) (Table 3.2). Therefore, psychotic symptoms and cognitive dysfunctions, observed during therapy with COX-1 inhibitors, were assigned to the COX-1-mediated increase of KYNA. The reduction of KYNA levels—by a prostaglandin-mediated mechanism—might be an additional mechanism to the above described immunological mechanism for therapeutic effects of selective COX-2 inhibitors in schizophrenia (Schwieler et al., 2005).

Indeed, in a prospective, randomized, double-blind study of therapy with the COX-2 inhibitor celecoxib add-on to risperidone in acute exacerbation of schizophrenia, a therapeutic effect of celecoxib was observed (Müller et al., 2002). Immunologically, an increase of the type-1 immune response was found in the celecoxib treatment group (Müller, Riedel, et al., 2004). The finding of a clinical advantage of COX-2 inhibition, however, could not be replicated in a second study. Further analysis of the data revealed that the outcome depends on the duration of the disease (Müller, Ulmschneider, et al., 2004). This observation is in accordance with results from animal studies showing that the effects of COX-2 inhibition on cytokines, hormones, and particularly on behavioral symptoms are dependent on the duration of the preceding changes and the time point of application of the COX-2 inhibitor (Casolini, Catalani, Zuena, & Angelucci, 2002). In subsequent clinical studies following a similar randomized, double-blind, placebo-controlled add-on design of 400 mg celecoxib to risperidone (in one study risperidone or olanzapine) in partly different patient populations, similar positive results of cyclooxygenase inhibition could be obtained: in a

Table 3.2 Overview on clinical studies with COX-2 inhibitors in schizophrenia

Authors	Diagnosis	Course and duration	Duration of trial	*N*	Study design	Concomitant drug	COX-2 inhibitor	Outcome
Zhang, Chun Chen, Long Tan, and Zhou (2006)	Schizophrenia	First manifestation	12 Weeks	40	Double-blind, randomized, placebo-controlled add-on	Risperidone (flexible dose)	Celecoxib (400 mg/day)	Significant advantage of the COX-2 inhibitor
Müller et al. (2010)	Schizophrenia	First manifestation	6 Weeks	49	Double-blind, randomized, placebo-controlled add-on	Amisulpride (flexible dose)	Celecoxib (400 mg/day)	Significant advantage of the COX-2 inhibitor
Müller et al. (2002)	Schizophrenia	Not specified (mean 5.9 years)	5 Weeks	50	Double-blind, randomized, placebo-controlled add-on	Risperidone (flexible dose)	Celecoxib (400 mg/day)	Significant advantage of the COX-2 inhibitor
Rappard and Müller (2004)	Schizophrenia	$\leq$10 Years	11 Weeks	270	Double-blind, randomized, placebo-controlled add-on	Risperidone (flexible dose)	Celecoxib (400 mg/day)	Non advantage of the COX-2 inhibitor
Rapaport et al. (2005)	Schizophrenia	Continuously ill (mean 20 years)	8 Weeks	38	Double-blind, randomized, placebo-controlled add-on	Risperidone or olanzapine (constant dose)	Celecoxib (400 mg/day)	Non-advantage on the COX-2 inhibitor
Akhondzadeh et al. (2007)	Schizophrenia	Chronic type (active phase)	8 Weeks	60	Double-blind, randomized, placebo-controlled add-on	Risperidone (fixed dose)	Celecoxib (400 mg/day)	Significant advantage of the COX-2 inhibitor

Chinese population of first manifestation schizophrenics (Zhang et al., 2006) and in an Iranian sample of chronic schizophrenics (Akhondzadeh et al., 2007). In continuously ill schizophrenics, however, no advantage of celecoxib could be found (Rapaport et al., 2005). A recent 6-week randomized, double-blind study using celecoxib add-on to amisulpride in first manifestation schizophrenia underlines the effect of short-term treatment with a COX-2 inhibitor in early stages of schizophrenia: the celecoxib group showed a significant better outcome not only on the PANS-total score but also on the PANS-negative and PANS global score (Müller et al., 2010). The PANS global score, in part, reflects the cognitive function of the schizophrenic patients.

Effects of COX-2 inhibitors on cognition (Müller, Riedel, Schwarz, & Engel, 2005) and on general psychopathology (Akhondzadeh et al., 2007) have been described before. An effect on cognition could also have been expected from the animal data of COX-2 inhibitors: COX-2 inhibition directly attenuates inflammation-induced inhibition of long-term potentiation, an animal model of cognition (Cumiskey, Curran, Herron, & O'Connor, 2007; Müller, Riedel, et al., 2005). Animals with a genetic overexpression of COX-2 showed more prominent deficits in cognition, which were attenuated by a selective COX-2 inhibitor (Melnikova et al., 2006).

In schizophrenia, COX-2 inhibition showed beneficial effects preferentially in early stages of the disease, the data regarding chronic schizophrenia are controversial, possibly in part due to methodological concerns. The data are still preliminary, and further research has to be performed, for example, with other COX-2 inhibitors.

ACKNOWLEDGMENT

Disclosures: Parts of this chapter (describing general aspects of immunological mechanisms in schizophrenia) have been published before.

REFERENCES

Ahuja, N., & Carroll, B. T. (2007). Possible anti-catatonic effects of minocycline in patients with schizophrenia. *Progress in Neuro-Psychopharmacology & Biological Psychiatry*, *31*, 968–969.

Akhondzadeh, S., Tabatabaee, M., Amini, H., Ahmadi Abhari, S. A., Abbasi, S. H., & Behnam, V. (2007). Celecoxib as adjunctive therapy in schizophrenia: A double-blind, randomized and placebo-controlled trial. *Schizophrenia Research*, *90*, 179–185.

Amminger, G. P., Schafer, M. R., Papageorgiou, K., Klier, C. M., Cotton, S. M., Harrigan, S. M., et al. (2010). Long-chain omega-3 fatty acids for indicated prevention of psychotic disorders: A randomized, placebo-controlled trial. *Archives of General Psychiatry*, *67*, 146–154.

Bayer, T. A., Buslei, R., Havas, L., & Falkai, P. (1999). Evidence for activation of microglia in patients with psychiatric illnesses. *Neuroscience Letters, 271*, 126–128.
Bechter, K. (2001). Mild encephalitis underlying psychiatric disorders—A reconsideration and hypothesis exemplified on Borna disease. *Neurology, Psychiatry and Brain Research, 9*, 55–70.
Bentsen, H. (2006). The Norwegian study on the treatment of schizophrenia and schizoaffective disorder with ethyl-EPA and antioxidants. *Presented at second conference on Brain Phospholipids*, Scotland: Aviemore.
Berger, G. E. (2004). Ethyl-eicosapentaenoic acid (E-EPA) supplementation in early psychosis: A double-blind randomised placebo-controlled add on study in 80 drug-naive first episode psychosis patients. *The International Journal of Neuropsychopharmacology, 8*, S422.
Bernstein, H. G., Steiner, J., & Bogerts, B. (2009). Glial cells in schizophrenia: Pathophysiological significance and possible consequences for therapy. *Expert Review of Neurotherapeutics, 9*, 1059–1071.
Block, M. L., Zecca, L., & Hong, J. S. (2007). Microglia-mediated neurotoxicity: Uncovering the molecular mechanisms. *Nature Reviews. Neuroscience, 8*, 57–69.
Bonartsev, P. D. (2008). An electron microscopic study of typical lymphocytes and atypical cells of peripheral blood in patients with schizophrenia. *Zhurnal Nevrologii i Psikhiatrii Imeni S.S. Korsakova, 108*, 62–68.
Bresee, C., & Rapaport, M. H. (2009). Persistently increased serum soluble interleukin-2 receptors in continuously ill patients with schizophrenia. *The International Journal of Neuropsychopharmacology, 12*, 861–865.
Brown, A. S. (2006). Prenatal infection as a risk factor for schizophrenia. *Schizophrenia Bulletin, 32*, 200–202.
Brown, A. S. (2008). The risk for schizophrenia from childhood and adult infections. *The American Journal of Psychiatry, 165*, 7–10.
Brown, A. S., Begg, M. D., Gravenstein, S., Schaefer, C. A., Wyatt, R. J., Bresnahn, M., et al. (2004). Serologic evidence of prenatal influenza in the etiology of schizophrenia. *Archives of General Psychiatry, 61*, 774–780.
Buka, S. L., Goldstein, J. M., Seidman, L. J., & Tsuang, M. T. (2000). Maternal recall of pregnancy history: Accuracy and bias in schizophrenia research. *Schizophrenia Bulletin, 26*, 335–350.
Casolini, P., Catalani, A., Zuena, A. R., & Angelucci, L. (2002). Inhibition of COX-2 reduces the age-dependent increase of hippocampal inflammatory markers, corticosterone secretion, and behavioral impairments in the rat. *Journal of Neuroscience Research, 68*, 337–343.
Cazzullo, C. L., Scarone, S., Grassi, B., Vismara, C., Trabattoni, D., Clerici, M., et al. (1998). Cytokines production in chronic schizophrenia patients with or without paranoid behaviour. *Progress in Neuro-Psychopharmacology & Biological Psychiatry, 22*, 947–957.
Chaves, C., Marque, C. R., Chaudhry, I. B., Hussain, F. M., Oliveira, J. P., & Wichert-Ana, L. (2009). Short-term improvement by minocycline added to olanzapine antipsychotic treatment in paranoid schizophrenia. *Schizophrenia Bulletin, 354* (Abstract).
Chiarugi, A., Carpenedo, R., & Moroni, F. (1996). Kynurenine disposition in blood and brain of mice: Effects of selective inhibitors of kynurenine hydroxylase and of kynureninase. *Journal of Neurochemistry, 67*, 692–698.
Chittiprol, S., Venkatasubramanian, G., Neelakantachar, N., Babu, S. V., Reddy, N. A., Shetty, K. T., et al. (2010). Oxidative stress and neopterin abnormalities in schizophrenia: A longitudinal study. *Journal of Psychiatric Research, 44*, 310–313.
Costa, E., Dong, E., Grayson, D. R., Guidotti, A., Ruzicka, W., & Veldic, M. (2007). Reviewing the role of DNA (cytosine-5) methyltransferase overexpression in the cortical GABAergic dysfunction associated with psychosis vulnerability. *Epigenetics, 2*, 29–36.

Cumiskey, D., Curran, B. P., Herron, C. E., & O'Connor, J. J. (2007). A role for inflammatory mediators in the IL-18 mediated attenuation of LTP in the rat dentate gyrus. *Neuropharmacology, 52,* 1616–1623.

Dalman, C., Allebeck, P., Gunnell, D., Harrison, G., Kristensson, K., Lewis, G., et al. (2008). Infections in the CNS during childhood and the risk of subsequent psychotic illness: A cohort study of more than one million Swedish subjects. *The American Journal of Psychiatry, 165,* 59–65.

Dameshek, W. (1930). White blood cells in dementia praecox and dementia paralytietwa. *Archives of Neurology and Psychiatry, 24,* 855.

Dickerson, F. B., Boronow, J. J., Stallings, C. R., Origoni, A. E., & Yolken, R. H. (2003). Reduction of symptoms by valacyclovir in cytomegalovirus-seropositive individuals with schizophrenia. *The American Journal of Psychiatry, 160,* 2234–2236.

Dickerson, F. B., Stallings, C. R., Boronow, J. J., Origoni, A. E., Sullens, A., & Yolken, R. H. (2009). Double blind trial of adjunctive valacyclovir in individuals with schizophrenia who are seropositive for cytomegalovirus. *Schizophrenia Research, 107,* 147–149.

Dickerson, F. B., Stallings, C. R., Boronow, J. J., Origoni, A. E., & Yolken, R. H. (2009). A double-blind trial of adjunctive azithromycin in individuals with schizophrenia who are seropositive for Toxoplasma gondii. *Schizophrenia Research, 112,* 198–199.

Doorduin, J., de Vries, E. F., Willemsen, A. T., de Groot, J. C., Dierckx, R. A., & Klein, H. C. (2009). Neuroinflammation in schizophrenia-related psychosis: A PET study. *Journal of Nuclear Medicine, 50,* 1801–1807.

Downes, C. E., & Crack, P. J. (2010). Neural injury following stroke: Are Toll-like receptors the link between the immune system and the CNS? *British Journal of Pharmacology, 160,* 1872–1888.

Drexhage, R. C., Knijff, E. M., Padmos, R. C., Heul-Nieuwenhuijzen, L., Beumer, W., Versnel, M. A., et al. (2010). The mononuclear phagocyte system and its cytokine inflammatory networks in schizophrenia and bipolar disorder. *Expert Review of Neurotherapeutics, 10,* 59–76.

Ehrenreich, H., Hinze-Selch, D., Stawicki, S., Aust, C., Knolle-Veentjer, S., Wilms, S., et al. (2007). Improvement of cognitive functions in chronic schizophrenic patients by recombinant human erythropoietin. *Molecular Psychiatry, 12,* 206–220.

Emsley, R., Myburgh, C., Oosthuizen, P., & van Rensburg, S. J. (2002). Randomized, placebo-controlled study of ethyl-eicosapentaenoic acid as supplemental treatment in schizophrenia. *The American Journal of Psychiatry, 159,* 1596–1598.

Emsley, R., Niehaus, D. J., Koen, L., Oosthuizen, P. P., Turner, H. J., Carey, P., et al. (2006). The effects of eicosapentaenoic acid in tardive dyskinesia: A randomized, placebo-controlled trial. *Schizophrenia Research, 84,* 112–120.

Fellerhoff, B., Laumbacher, B., Mueller, N., Gu, S., & Wank, R. (2007). Associations between Chlamydophila infections, schizophrenia and risk of HLA-A10. *Molecular Psychiatry, 12,* 264–272.

Fenton, W. S., Dickerson, F., Boronow, J., Hibbeln, J. R., & Knable, M. (2001). A placebo-controlled trial of omega-3 fatty acid (ethyl eicosapentaenoic acid) supplementation for residual symptoms and cognitive impairment in schizophrenia. *The American Journal of Psychiatry, 158,* 2071–2074.

Gattaz, W. F., Abrahao, A. L., & Foccacia, R. (2004). Childhood meningitis, brain maturation and the risk of psychosis. *European Archives of Psychiatry and Clinical Neuroscience, 254,* 23–26.

Guillemin, G. J., Kerr, S. J., Smythe, G. A., Smith, D. G., Kapoor, V., Armati, P. J., et al. (2001). Kynurenine pathway metabolism in human astrocytes: A paradox for neuronal protection. *Journal of Neurochemistry, 78,* 842–853.

Haack, M., Hinze-Selch, D., Fenzel, T., Kraus, T., Kuhn, M., Schuld, A., et al. (1999). Plasma levels of cytokines and soluble cytokine receptors in psychiatric patients upon hospital admission: Effects of confounding factors and diagnosis. *Journal of Psychiatric Research*, *33*, 407–418.

Hanisch, U. K., & Kettenmann, H. (2007). Microglia: Active sensor and versatile effector cells in the normal and pathologic brain. *Nature Neuroscience*, *10*, 1387–1394.

Heyes, M. P., Chen, C. Y., Major, E. O., & Saito, K. (1997). Different kynurenine pathway enzymes limit quinolinic acid formation by various human cell types. *The Biochemical Journal*, *326*, 351–356.

Jack, C. S., Arbour, N., Manusow, J., Montgrain, V., Blain, M., McCrea, E., et al. (2005). TLR signaling tailors innate immune responses in human microglia and astrocytes. *Journal of Immunology (Baltimore, Md.: 1950)*, *175*, 4320–4330.

Kawasaki, H., & Iwamuro, S. (2008). Potential roles of histones in host defense as antimicrobial agents. *Infectious Disorders Drug Targets*, *8*, 195–205.

Kiss, C., Ceresoli-Borroni, G., Guidetti, P., Zielke, C. L., Zielke, H. R., & Schwarcz, R. (2003). Kynurenate production by cultured human astrocytes. *Journal of Neural Transmission*, *110*, 1–14.

Koponen, H., Rantakallio, P., Veijola, J., Jones, P., Jokelainen, J., & Isohanni, M. (2004). Childhood central nervous system infections and risk for schizophrenia. *European Archives of Psychiatry and Clinical Neuroscience*, *254*, 9–13.

Körschenhausen, D. A., Hampel, H. J., Ackenheil, M., Penning, R., & Müller, N. (1996). Fibrin degradation products in post mortem brain tissue of schizophrenics: A possible marker for underlying inflammatory processes. *Schizophrenia Research*, *19*, 103–109.

Korte, S., Arolt, V., Peters, M., Weitzsch, C., Rothermundt, M., & Kirchner, H. (1998). Increased serum neopterin levels in acutely ill and recovered schizophrenic patients. *Schizophrenia Research*, *32*, 63–67.

Lin, A., Kenis, G., Bignotti, S., Tura, G. J., De Jong, R., Bosmans, E., et al. (1998). The inflammatory response system in treatment-resistant schizophrenia: Increased serum interleukin-6. *Schizophrenia Research*, *32*, 9–15.

Maes, M., Bosmans, E., De Jongh, R., Kenis, G., Vandoolaeghe, E., & Neels, H. (1997). Increased serum IL-6 and IL-1 receptor antagonist concentrations in major depression and treatment resistant depression. *Cytokine*, *9*, 853–858.

Melnikova, T., Savonenko, A., Wang, Q., Liang, X., Hand, T., Wu, L., et al. (2006). Cycloxygenase-2 activity promotes cognitive deficits but not increased amyloid burden in a model of Alzheimer's disease in a sex-dimorphic pattern. *Neuroscience*, *141*, 1149–1162.

Meyer, U., Schwarz, M. J., & Müller, N. (2011). Inflammatory processes in schizophrenia: A promising neuroimmunological target for the treatment of negative/cognitive symptoms and beyond. *Pharmacology & Therapeutics*, *132*, 96–110.

Mills, C. D., Kincaid, K., Alt, J. M., Heilman, M. J., & Hill, A. M. (2000). M-1/M-2 macrophages and the Th1/Th2 paradigm. *Journal of Immunology (Baltimore, Md.: 1950)*, *164*, 6166–6173.

Mittleman, B. B., Castellanos, F. X., Jacobsen, L. K., Rapoport, J. L., Swedo, S. E., & Shearer, G. M. (1997). Cerebrospinal fluid cytokines in pediatric neuropsychiatric disease. *Journal of Immunology (Baltimore, Md.: 1950)*, *159*, 2994–2999.

Mizoguchi, H., Takuma, K., Fukakusa, A., Ito, Y., Nakatani, A., Kim, H. C., et al. (2008). Improvement by minocycline of methamphetamine-induced impairment of recognition memory in mice. *Psychopharmacology*, *196*, 233–241.

Molholm, H. B. (1942). Hyposensitivity to foreign protein in schizophrenic patients. *Psychiatric Quarterly*, *16*, 565–571.

Müller, N., Ackenheil, M., Hofschuster, E., Mempel, W., & Eckstein, R. (1991). Cellular immunity in schizophrenic patients before and during neuroleptic treatment. *Psychiatry Research*, *37*, 147–160.

Müller, N., Empl, M., Riedel, M., Schwarz, M., & Ackenheil, M. (1997). Neuroleptic treatment increases soluble IL-2 receptors and decreases soluble IL-6 receptors in schizophrenia. *European Archives of Psychiatry and Clinical Neuroscience*, *247*, 308–313.

Müller, N., Krause, D., Dehning, S., Musil, R., Schennach-Wolff, R., Obermeier, M., et al. (2010). Celecoxib treatment in an early stage of schizophrenia: Results of a randomized, double-blind, placebo-controlled trial of celecoxib augmentation of amisulpride treatment. *Schizophrenia Research*, *121*, 119–124.

Müller, N., Riedel, M., Ackenheil, M., & Schwarz, M. J. (2000). Cellular and humoral immune system in schizophrenia: A conceptual re-evaluation. *The World Journal of Biological Psychiatry*, *1*, 173–179.

Müller, N., Riedel, M., Dehning, S., Spellmann, I., Müller-Arends, A., Cerovecki, A., et al. (2004). Is the therapeutic effect of celecoxib in schizophrenia depending from duration of disease? *Neuropsychopharmacology*, *29*, 176.

Müller, N., Riedel, M., Hadjamu, M., Schwarz, M. J., Ackenheil, M., & Gruber, R. (1999). Increase in expression of adhesion molecule receptors on T helper cells during antipsychotic treatment and relationship to blood-brain barrier permeability in schizophrenia. *The American Journal of Psychiatry*, *156*, 634–636.

Müller, N., Riedel, M., Scheppach, C., Brandstätter, B., Sokullu, S., Krampe, K., et al. (2002). Beneficial antipsychotic effects of celecoxib add-on therapy compared to risperidone alone in schizophrenia. *The American Journal of Psychiatry*, *159*, 1029–1034.

Müller, N., Riedel, M., Schwarz, M. J., & Engel, R. R. (2005). Clinical effects of COX-2 inhibitors on cognition in schizophrenia. *European Archives of Psychiatry and Clinical Neuroscience*, *255*, 149–151.

Müller, N., Riedel, M., Schwarz, M. J., Gruber, R., & Ackenheil, M. (1997). Immunomodulatory effects of neuroleptics to the cytokine system and the cellular immune system in schizophrenia. In G. Wieselmann (Ed.), *Current update in psychoimmunology* (pp. 57–67). Wien: Springer.

Müller, N., & Schwarz, M. J. (2007a). The immune-mediated alteration of serotonin and glutamate: Towards an integrated view of depression. *Molecular Psychiatry*, *12*, 988–1000.

Müller, N., & Schwarz, M. J. (2007b). The immunological basis of glutamatergic disturbance in schizophrenia: Towards an integrated view. *Journal of Neural Transmission. Supplementum*, *72*, 269–280.

Müller, N., & Schwarz, M. J. (2012). Immunological treatment options for schizophrenia. *Current Pharmaceutical Biotechnology*. Jan 26 (Epub ahead of print).

Müller, N., Schwarz, M. J., & Riedel, M. (2005). COX-2 inhibition in schizophrenia: Focus on clinical effects of celecoxib therapy and the role of TNF-alpha. In W. W. Eaton (Ed.), *Medical and psychiatric comorbidity over the course of life* (pp. 265–276). Washington, DC: American Psychiatric Publishing.

Müller, N., Ulmschneider, M., Scheppach, C., Schwarz, M. J., Ackenheil, M., Möller, H. J., et al. (2004). COX-2 inhibition as a treatment approach in schizophrenia: Immunological considerations and clinical effects of celecoxib add-on therapy. *European Archives of Psychiatry and Clinical Neuroscience*, *254*, 14–22.

Müller, N., Wagner, J. K., Krause, D., Weidinger, E., Wildenauer, A., Obermeier, M., et al. (2012). Impaired monocyte activation in schizophrenia. *Psychiatry Research*. Mar 17 (Epub ahead of print).

Na, K. S., & Kim, Y. K. (2007). Monocytic, Th1 and th2 cytokine alterations in the pathophysiology of schizophrenia. *Neuropsychobiology*, *56*, 55–63.

Nikkilä, H. V., Muller, K., Ahokas, A., Rimon, R., & Andersson, L. C. (2001). Increased frequency of activated lymphocytes in the cerebrospinal fluid of patients with acute schizophrenia. *Schizophrenia Research*, *49*, 99–105.

Ozek, M., Toreci, K., Akkok, I., & Guvener, Z. (1971). Influence of therapy on antibody-formation. *Psychopharmacologia, 21*, 401–412.
Peet, M. (2008). Omega-3 polyunsaturated fatty acids in the treatment of schizophrenia. *Israel Journal of Psychiatry & Related Sciences, 45*, 19–25.
Peet, M., Brind, J., Ramchand, C. N., Shah, S., & Vankar, G. K. (2001). Two double-blind placebo-controlled pilot studies of eicosapentaenoic acid in the treatment of schizophrenia. *Schizophrenia Research, 49*, 243–251.
Peet, M., & Horrobin, D. F. (2002). A dose-ranging exploratory study of the effects of ethyl-eicosapentaenoate in patients with persistent schizophrenic symptoms. *Journal of Psychiatric Research, 36*, 7–18.
Pollmächer, T., Schuld, A., Kraus, T., Haack, M., & Hinze-Selch, D. (2001). [On the clinical relevance of clozapine-triggered release of cytokines and soluble cytokine-receptors]. *Fortschritte der Neurologie-Psychiatrie, 69*(Suppl. 2), S65–S74.
Potvin, S., Stip, E., Sepehry, A. A., Gendron, A., Bah, R., & Kouassi, E. (2008). Inflammatory cytokine alterations in schizophrenia: A systematic quantitative review. *Biological Psychiatry, 63*, 801–808.
Purcell, S. M., Wray, N. R., Stone, J. L., Visscher, P. M., O'Donovan, M. C., Sullivan, P. F., et al. (2009). Common polygenic variation contributes to risk of schizophrenia and bipolar disorder. *Nature, 460*, 748–752.
Radewicz, K., Garey, L. J., Gentleman, S. M., & Reynolds, R. (2000). Increase in HLA-DR immunoreactive microglia in frontal and temporal cortex of chronic schizophrenics. *Journal of Neuropathology and Experimental Neurology, 59*, 137–150.
Rapaport, M. H., Delrahim, K. K., Bresee, C. J., Maddux, R. E., Ahmadpour, O., & Dolnak, D. (2005). Celecoxib augmentation of continuously ill patients with schizophrenia. *Biological Psychiatry, 57*, 1594–1596.
Rappard, F., & Müller, N. (2004). Celecoxib add-on therapy does not have beneficial antipsychotic effects over risperidone alone in schizophrenia. *Neuropsychopharmacology, 29* (Suppl. 1), S222.
Riedel, M., Spellmann, I., Schwarz, M. J., Strassnig, M., Sikorski, C., Möller, H. J., et al. (2007). Decreased T cellular immune response in schizophrenic patients. *Journal of Psychiatric Research, 41*, 3–7.
Ross, B. M., Seguin, J., & Sieswerda, L. E. (2007). Omega-3 fatty acids as treatments for mental illness: Which disorder and which fatty acid? *Lipids in Health and Disease, 6* (21), 21.
Rothermundt, M., Arolt, V., Weitzsch, C., Eckhoff, D., & Kirchner, H. (1998). Immunological dysfunction in schizophrenia: A systematic approach. *Neuropsychobiology, 37*, 186–193.
Rothermundt, M., Falkai, P., Ponath, G., Abel, S., Burkle, H., Diedrich, M., et al. (2004). Glial cell dysfunction in schizophrenia indicated by increased S100B in the CSF. *Molecular Psychiatry, 9*, 897–899.
Schwarz, M. J., Chiang, S., Müller, N., & Ackenheil, M. (2001). T-helper-1 and T-helper-2 responses in psychiatric disorders. *Brain, Behavior, and Immunity, 15*, 340–370.
Schwarz, M. J., Riedel, M., Ackenheil, M., & Müller, N. (2000). Decreased levels of soluble intercellular adhesion molecule-1 (sICAM-1) in unmedicated and medicated schizophrenic patients. *Biological Psychiatry, 47*, 29–33.
Schwieler, L., Erhardt, S., Erhardt, C., & Engberg, G. (2005). Prostaglandin-mediated control of rat brain kynurenic acid synthesis—Opposite actions by COX-1 and COX-2 isoforms. *Journal of Neural Transmission, 112*, 863–872.
Shi, J., Levinson, D. F., Duan, J., Sanders, A. R., Zheng, Y., Pe'er, I., et al. (2009). Common variants on chromosome 6p22.1 are associated with schizophrenia. *Nature, 460*, 753–757.
Sperner-Unterweger, B., Miller, C., Holzner, B., Widner, B., Fleischhacker, W. W., & Fuchs, D. (1999). Measurement of neopterin, kynurenine and tryptophan in sera of

schizophrenic patients. In N. Müller (Ed.), *Psychiatry, psychoimmunology, and viruses* (pp. 115–119). Wien: Springer.

Sperner-Unterweger, B., Whitworth, A., Kemmler, G., Hilbe, W., Thaler, J., Weiss, G., et al. (1999). T-cell subsets in schizophrenia: A comparison between drug-naive first episode patients and chronic schizophrenic patients. *Schizophrenia Research, 38*, 61–70.

Stefansson, H., Ophoff, R. A., Steinberg, S., Andreassen, O. A., Cichon, S., Rujescu, D., et al. (2009). Common variants conferring risk of schizophrenia. *Nature, 460*, 744–747.

Steiner, J., Bielau, H., Brisch, R., Danos, P., Ullrich, O., Mawrin, C., et al. (2008). Immunological aspects in the neurobiology of suicide: Elevated microglial density in schizophrenia and depression is associated with suicide. *Journal of Psychiatric Research, 42*, 151–157.

Swerdlow, N. R., van Bergeijk, D. P., Bergsma, F., Weber, E., & Talledo, J. (2009). The effects of memantine on prepulse inhibition. *Neuropsychopharmacology, 34*, 1854–1864.

Takano, A., Arakawa, R., Ito, H., Tateno, A., Takahashi, H., Matsumoto, R., et al. (2010). Peripheral benzodiazepine receptors in patients with chronic schizophrenia: A PET study with [11C]DAA1106. *The International Journal of Neuropsychopharmacology, 13*, 943–950.

Tanaka, K. F., Shintani, F., Fujii, Y., Yagi, G., & Asai, M. (2000). Serum interleukin-18 levels are elevated in schizophrenia. *Psychiatry Research, 96*, 75–80.

van Berckel, B. N., Bossong, M. G., Boellaard, R., Kloet, R., Schuitemaker, A., Caspers, E., et al. (2008). Microglia activation in recent-onset schizophrenia: A quantitative (R)-[11C]PK11195 positron emission tomography study. *Biological Psychiatry, 64*, 820–822.

van Kammen, D. P., McAllister-Sistilli, C. G., & Kelley, M. E. (1997). Relationship between immune and behavioral measures in schizophrenia. In G. Wieselmann (Ed.), *Current update in psychoimmunology* (pp. 51–55). Wien: Springer.

Wagner von Jauregg, J. (1926). Fieberbehandlung bei Psychosen. *Wiener Medizinische Wochenschrift, 76*, 79–82.

Wierzba-Bobrowicz, T., Lewandowska, E., Lechowicz, W., Stepien, T., & Pasennik, E. (2005). Quantitative analysis of activated microglia, ramified and damage of processes in the frontal and temporal lobes of chronic schizophrenics. *Folia Neuropathologica, 43*, 81–89.

Wilke, I., Arolt, V., Rothermundt, M., Weitzsch, C., Hornberg, M., & Kirchner, H. (1996). Investigations of cytokine production in whole blood cultures of paranoid and residual schizophrenic patients. *European Archives of Psychiatry and Clinical Neuroscience, 246*, 279–284.

Winter, C., Djodari-Irani, A., Sohr, R., Morgenstern, R., Feldon, J., Juckel, G., et al. (2009). Prenatal immune activation leads to multiple changes in basal neurotransmitter levels in the adult brain: Implications for brain disorders of neurodevelopmental origin such as schizophrenia. *The International Journal of Neuropsychopharmacology, 12*, 513–524.

Wüstenberg, T., Begemann, M., Bartels, C., Gefeller, O., Stawicki, S., Hinze-Selch, D., et al. (2011). Recombinant human erythropoietin delays loss of gray matter in chronic schizophrenia. *Molecular Psychiatry, 16*, 26–36.

Zhang, Y., Chun Chen, D., Long Tan, Y., & Zhou, D. F. (2006). A double-blind, placebo-controlled trial of celecoxib addes to risperidone in first-episode and drug-naive patients with schizophrenia. *European Archives of Psychiatry and Clinical Neuroscience, 256*(Suppl. 2), II/50.

Zuckerman, L., & Weiner, I. (2005). Maternal immune activation leads to behavioral and pharmacological changes in the adult offspring. *Journal of Psychiatric Research, 39*, 311–323.

CHAPTER FOUR

Inflammation in Parkinson's Disease

Kemal Ugur Tufekci*, Ralph Meuwissen[†], Sermin Genc*, Kursad Genc*[1]

*Department of Neuroscience, Health Science Institute, Dokuz Eylul University, Izmir, Turkey
[†]Department of Internal Medicine, School of Medicine, Dokuz Eylul University, Izmir, Turkey
[1]Corresponding author: e-mail address: kkursadgenc@hotmail.com

Contents

Advances in Protein Chemistry and Structural Biology, Volume 88
ISSN 1876-1623
http://dx.doi.org/10.1016/B978-0-12-398314-5.00004-0

Abstract

Parkinson's disease (PD) is a common neurodegenerative disease that is characterized by the degeneration of dopaminergic neurons in the substantia nigra pars compacta. Inflammatory responses manifested by glial reactions, T cell infiltration, and increased expression of inflammatory cytokines, as well as other toxic mediators derived from activated glial cells, are currently recognized as prominent features of PD. The consistent findings obtained by various animal models of PD suggest that neuroinflammation is an important contributor to the pathogenesis of the disease and may further propel the progressive loss of nigral dopaminergic neurons. Furthermore, although it may not be the primary cause of PD, additional epidemiological, genetic, pharmacological, and imaging evidence support the proposal that inflammatory processes in this specific brain region are crucial for disease progression. Recent *in vitro* studies, however, have suggested that activation of microglia and subsequently astrocytes via mediators released by injured dopaminergic neurons is involved. However, additional *in vivo* experiments are needed for a deeper understanding of the mechanisms involved in PD pathogenesis. Further insight on the mechanisms of inflammation in PD will help to further develop alternative therapeutic strategies that will specifically and temporally target inflammatory processes without abrogating the potential benefits derived by neuroinflammation, such as tissue restoration.

ABBREVIATIONS

6-OHDA 6-hydroxydopamine
AD Alzheimer's disease
AICD activation-induced cell death
ALP autophagy/lysosomal degradation pathway
ALS amyotrophic lateral sclerosis
AMPA 2-amino-3-(5-methyl-3-oxo-1,2-oxazol-4-yl)propanoic acid
AP-1 activator protein-1
APC antigen-presenting cell
ARE antioxidant response element
AR-JP autosomal recessive juvenile-PD
AR-PD autosomal recessive-PD
AT1 angiotensin II type 1 receptor
AT2 angiotensin II type 2 receptor
ATP13A2 ATPase type 13A2
Bad Bcl-2-associated agonist of cell death
Bax Bcl-2-associated X protein
BBB blood–brain barrier

BCG Bacille Calmette–Guérin
Bcl-2 B-cell lymphoma 2
BDNF brain-derived neurotrophic factor
BG basal ganglia
Bim Bcl2-like 11
C1qA complement C1q subcomponent subunit A
CatD cathepsin D
CDNF cerebral dopamine neurotrophic factor
CMA chaperone-mediated autophagy
CNS central nervous system
CNV copy number variation
COMT catechol-*O*-methyl transferase
COMTI catechol-*O*-methyl transferase inhibitor
COR C-terminal of ROC
COX2 cyclooxygenase-2
CSF cerebrospinal fluid
CTLA-4 cytotoxic T lymphocyte antigen 4
CXCL14 chemokine (C-X-C motif) ligand 14
CYP1A2 cytochrome P450 1A2
DAMP danger-associated molecular pattern
DAT dopamine transporter
DBS deep brain stimulation
DJ-1 Daisuke-Junko-1
EIF4G1 eukaryotic translation initiation factor 4 gamma, 1
FADD Fas-associated death domain protein
FBXO7 F-box protein 7
GA glatiramer acetate
GABA γ-aminobutyric acid
GAK cyclin G-associated kinase
GBA glucocerebrosidase
GDNF glial cell-derived neurotrophic factor
GFAP glial fibrillary acidic protein
GPe globus pallidus pars externa
GPi globus pallidus pars interna
Gpx glutathione peroxidase
GR glucocorticoid receptor
GRIN2A glutamate receptor subunit epsilon-1
GWAS genome-wide association studies
HD Huntington's disease
HIV human immunodeficiency virus
HLA-DR human leukocyte antigen DR
HMGB-1 high mobility group box chromosomal protein 1
HSP73 heat-shock protein 73
ICAM-1 intercellular cell adhesion molecule-1
IFN-γ interferon-γ
Ig immunoglobulin
IκB inhibitor of kappa B

IL-10 interleukin-10
IL-12 interleukin-12
IL-13 interleukin-13
IL-14 interleukin-14
IL-18 interleukin-18
IL-1RA IL-1 receptor antagonist
IL-1α interleukin-1α
IL-1β interleukin-1β
IL-2 interleukin-2
IL-6 interleukin-6
IL-7 interleukin-7
IL-8 interleukin-8
iLBD incidental Lewy body disease
iNOS inducible nitric oxide synthase
IP_3 inositol triphosphate
JE Japanese encephalitis virus
JIP JNK-interacting protein
JNK c-jun N-terminal kinase
LAMP1 lysosome-associated membrane protein 1
LB Lewy body
LC locus coeruleus
L-DOPA levodopa
LN Lewy neurite
LPS lipopolysaccharide
LRRK2 leucine-rich repeat kinase 2
MAC cytolytic membrane attack complexes
MAO-B monoamine oxidase type B
MAPK mitogen-activated protein kinase
MAPT microtubule-associated protein tau
MHC major histocompatibility complex
MK2 MAPK-activated protein kinase 2
MMP-3 matrix metalloproteinase-3
MPO myeloperoxidase
MPP^+ 1-methyl-4-phenylpyridinium
MPTP 1-methyl-4-phenyl-1,2,3,6-tetrahydropyridine
MRC1 mannose receptor c-type 1
MSA multiple system atrophy
mtDNA mitochondrial DNA
MyD88 myeloid differentiation response gene 88
NADPH nicotinamide adenine dinucleotide phosphate (reduced)
NFκB nuclear factor kappa-light-chain-enhancer of activated B cells
NKT natural killer T cell
NLR nucleotide oligomerization domain receptor
NMDA *N*-methyl-D-aspartic acid
NO nitric oxide
Nrf2 nuclear factor E2-related factor 2

NSAID nonsteroidal anti-inflammatory drug
Nurr1 nuclear receptor-related 1
NVU neurovascular unit
PAMP pathogen-associated molecular pattern
PB pale bodies
PBMC peripheral blood mononuclear cell
PD Parkinson's disease
PDGFB platelet-derived growth factor subunit B
PET positron emission tomography
PGE2 prostaglandin E2
PGJ2 prostaglandin J2
P-gp P-glycoprotein
PINK1 phosphatase and tensin homolog (PTEN)-induced putative kinase 1
PLA2G6 cytosolic, calcium-independent phospholipase A2, group VI
poly(I:C) polyinosinic:polycytidylic acid
PPARγ peroxisome proliferator-activated receptor γ
PQ paraquat
PRR pattern recognition receptor
PSP progressive supranuclear palsy
RAS renin–angiotensin system
Rep1 repeat polymorphism
RNS reactive nitrogen species
ROC ras-of-complex
ROS reactive oxygen species
SNpc substantia nigra pars compacta
SNr substantia nigra pars reticulata
STN subthalamic nucleus
T_C cytotoxic T cells
T_{eff} effector T cell
T_{fam} mitochondrial transcription factor A
TGF-β transforming growth factor-β
T_H helper T cells
TH tyrosine hydroxylase
TLR Toll-like receptor
TNF tumor necrosis factor
TRADD TNFR1-associated death domain adaptor protein
TRAF6 TNF-associated factor 6
T_{reg} regulatory T cells
TREM2 triggering receptor expressed on myeloid cells 2
UCH-L1 ubiquitin C-terminal hydrolase
UPR unfolded protein response
UPS ubiquitin-proteasomal system
VIP vasoactive intestinal protein
VMAT2 vesicular monoamine transporter 2
VPS35 vacuolar protein sorting 35 homolog
α-SYN α-synuclein

1. INTRODUCTION

The central nervous system (CNS) is considered to be a relatively immune-privileged tissue due to several factors including the absence of dendritic cells and the presence of an immunosuppressant microenvironment in the brain parenchyma under physiological conditions, the existence of a blood–brain barrier (BBB) that separates the brain parenchyma and the immune system in the periphery, and limited lymphatic drainage from the brain (Galea, Bechmann, & Perry, 2007). However, this does not mean that the CNS cannot initiate any immune responses against various insults such as pathogens or endogenous danger signals. Microglia, the resident tissue macrophages of the CNS, have the ability to initiate innate immune responses upon activation by various stimuli (Saijo & Glass, 2011; Tansey & Goldberg, 2010). Immune responses are then amplified after receiving signals from the astrocytes (Halliday & Stevens, 2011; Saijo & Glass, 2011). To maintain proper tissue homeostasis and to avoid collateral tissue damage during inflammatory reactions, all of the inflammatory responses must be resolved and terminated in the CNS similar to the peripheral immune system (Glass, Saijo, Winner, Marchetto, & Gage, 2010). This entails the rigorous completion of inflammatory tasks such as the removal of pathogens, dead cells, or other cellular debris and tissue repair. However, if the insult, regardless of its nature, still persists or if the resolution mechanisms for the inflammation are inadequate despite the lack of a triggering stimulus, then chronic inflammation can arise.

In the past decade, a growing interest in the inflammatory processes in age-related neurodegenerative diseases, including Parkinson's disease (PD), has emerged. Inflammation in the CNS is a prominent and common feature of such diseases, although the presence of several disease-specific inflammatory reactions remains elusive (Glass et al., 2010). Each of these disorders is characterized by a region-specific loss of specific neuronal subpopulations (such as the nigral dopaminergic neurons in PD) and extracellular [amyloid-β in Alzheimer's disease (AD)] or intracellular [α-synuclein (α-SYN) in PD] accumulation of specific disease proteins. Furthermore, the inflammatory responses induced by soluble factors secreted from injured neurons under neurodegenerative conditions are collectively known as neuroinflammation, which is widely regarded as a crucial player in the neurodegenerative process (Tansey & Goldberg, 2010). However, whether neuroinflammation is a causal/trigger factor or a secondary consequence in neurodegenerative diseases remains unknown (Graeber, Li, & Rodriguez, 2011).

PD is the second most common age-related neurodegenerative disease and the most common movement disorder and is characterized by a gradual and progressive loss of dopaminergic neurons in the substantia nigra pars compacta (SNpc). In this multifactorial, complex disorder, several converging mechanisms, including protein mishandling, mitochondrial dysfunction, oxidative stress, impaired autophagy, glutamate excitotoxicity, and neuroinflammation, likely play roles in the disease pathogenesis (Blandini, 2010; Valente, Arena, Torosantucci, & Gelmetti, 2012). Genetic factors act in concert with environmental influences, and epigenetic factors may be crucial in the initiation and acceleration of the disease process. As evidenced by experimental studies in animal models of the disease, neuroinflammation has been shown to be an important contributor (if not an initiating factor) to PD progression (Jackson-Lewis, Blesa, & Przedborski, 2012; Phani, Loike, & Przedborski, 2012; Tansey & Goldberg, 2010). A complex interplay appears between neuroinflammation and other proposed pathogenic mechanisms of PD such as mitochondrial dysfunction and oxidative stress (Witte, Geurts, de Vries, van der Valk, & van Horssen, 2010). Furthermore, recent studies suggest the involvement of protein products of parkinsonian genes (i.e., *α-SYN*, Parkin, and *DJ-1*) in innate immune responses (Beraud & Maguire-Zeiss, 2012; Cornejo Castro et al., 2010; Gardet et al., 2010; Greene, Whitworth, Andrews, Parker, & Pallanck, 2005; Hakimi et al., 2011; Roodveldt, Christodoulou, & Dobson, 2008; Thevenet, Pescini Gobert, Hooft van Huijsduijnen, Wiessner, & Sagot, 2011). Recently, an immunoregulatory role for the neurotransmitter dopamine has emerged (Sarkar, Basu, Chakroborty, Dasgupta, & Basu, 2010). Moreover, inflammatory stimuli can induce the expression of Parkin (Tran et al., 2011). Inflammatory responses may also contribute to the intrinsic vulnerability of nigral dopaminergic neurons, which is currently explained by several factors including pacemaking activity, dopaminergic metabolism, high iron content, a differential transcriptional profile, diverse calcium channel expression, and a relatively low antioxidant defense system (Double, 2012). Indeed, the nigral microglial density is highest in the rodent brain and is fairly high in the human brain (Kim et al., 2000; Lawson, Perry, Dri, & Gordon, 1990; Mittelbronn, Dietz, Schluesener, & Meyermann, 2001).

Neuroinflammatory responses in PD consist of phenotypic and morphological activation of glia cells and, to a lesser extent, include the recruitment of T cells from the peripheral immune system and increased expression and release of proinflammatory cytokines, chemokines, elements of the complement cascade, increased expression of several enzymes [NADPH oxidase, inducible nitric oxide synthase (iNOS), and myeloperoxidase (MPO)] that are involved in the excess production of reactive oxygen and nitrogen species and the increased expression of cyclooxygenase-2 (COX2), an enzyme involved in the synthesis of arachidonic acid derivatives such as prostaglandin E2 (PGE2) (Lull & Block, 2010; Tansey & Goldberg, 2010). Microglia, which constitute the main cellular components of neuroinflammation, are in a resting (but surveillant) state in the absence of any stimulus. This is accomplished by the existing immunosuppressant microenvironment in the CNS, where immunoregulatory molecules are expressed and/or released by healthy neurons (Glass et al., 2010). Thus, during neuronal death, the halting action on microglia is removed. Similarly, the limited function of astroglial cells on microglial activation also fails because of phenotypic changes in the astrocytes upon activation or potentially because of the excess activation of microglia that overcomes the braking signals derived from astrocytes and neurons (Saijo & Glass, 2011). Microglial overactivation may also affect the normal barrier function of cerebral endothelial cells, at least *in vitro* (Kacimi, Giffard, & Yenari, 2011). The interaction between microglia and cerebral endothelial cells is bidirectional (Wang, Li, et al., 2011). With the potential unfavorable contribution of dysfunctional astrocytes, molecular alterations may occur in the BBB, which permits the recruitment of T cells to the CNS and is promoted by chemotactic factors released by activated microglia. Thus, the physiological cross talk between CNS cells may be abrogated under inflammatory conditions. While proinflammatory molecules are in the repressed state under physiological conditions, upon activation, their increased levels act as toxic mediators of neuronal death in the neurodegenerative process. Furthermore, increased expression of some ligand–receptor pairs, such as the Fas/Fas ligand, has been implicated in neuronal death after cell-to-cell contact (Brochard et al., 2009). In the PD brain, death of nigral dopaminergic neurons occurs in an asynchronous and gradual fashion. Studies have shown that dying neurons release soluble mediators, such as α-SYN, matrix metalloproteinase-3 (MMP-3), neuromelanin, ATP, and μ-calpain, which cause microglia to secrete toxic mediators that are lethal to neighboring cells and stress neurons (Block, Zecca, & Hong, 2007; Lull & Block, 2010). Proinflammatory mediators released by activated glia act on their cognate receptors expressed on microglia and further increase the microglial activation status by rendering them to an overactivated state. This self-propelling vicious cycle leads to uncontrolled neuroinflammation that is deleterious to neurons but exerts some beneficial effects, including activity repair and restoration of tissue homeostasis (Lull & Block, 2010). Endogenous mechanisms mediate the resolution of inflammation, which include the actions of the transrepressor Nurr1 in glial cells. Endogenous glucocorticoids, microRNA-mediated posttranscriptional gene regulation or epigenetic mechanisms, may also fail to properly function during persistent inflammation (Alam & O'Neill, 2011; Hirsch, Vyas, & Hunot, 2012; McCall, El Gazzar, Liu, Vachharajani, & Yoza, 2011; Saijo et al., 2009). Currently, neuronal death is not solely regarded as a neuronal process in neurodegenerative diseases. Dysfunction of glial cells might result from the accumulation of misfolded, modified, and mutant disease proteins, which abolishes the trophic support to neurons that is otherwise provided by healthy glial cells via neurotrophic factors under physiological conditions (Halliday & Stevens, 2011). The disruption of the glial-antioxidant defense system causes stressed neurons to be additionally vulnerable to oxidative stress (Halliday & Stevens, 2011). Dysfunctional glia may also contribute to glutamate excitotoxicity, one of the proposed pathogenic mechanisms of PD, via the breakdown of glutamate release and uptake mechanisms achieved by normal glial cells (Blandini, 2010; Sofroniew & Vinters, 2010). Future human studies, *in vivo* animal model studies, and *in vitro* experiments will help us to clarify the underlying mechanisms of neuroinflammation in PD.

In this chapter, we review the accumulating epidemiological, genetic, pharmacological, and imaging evidence from human studies that suggest the involvement of inflammatory processes in PD pathogenesis. In addition, experimental studies that shed light on the putative mechanisms of neuroinflammation-mediated neurodegeneration are summarized. Lastly, the current and future statuses of anti-inflammatory treatments and immunization strategies that enable the specific and temporal manipulation of inflammatory processes during neurodegeneration are also evaluated.

2. PARKINSON'S DISEASE

2.1. Epidemiology

PD is the most common movement disorder and the second most frequent neurodegenerative disease after AD. The incidence rate of PD ranges between 1.5 and 22 in 100,000 person-years for all age groups (Wirdefeldt, Adami, Cole, Trichopoulos, & Mandel, 2011). The onset of disease commonly occurs during the sixth decade, but there are young (20–40 years) and juvenile onset (under 21 years) forms of PD (Muthane et al., 1994). Several studies have reported that the prevalence of PD ranges between 31 and 970 per 100,000 people (de Lau & Breteler, 2006; Wirdefeldt et al., 2011). The projection of European prevalence study results of 2030 shows an estimated doubling in the number of patients with PD compared to 2005 results (Dorsey et al., 2007).

2.2. Risk and protective factors

Age is a major risk factor for PD; which the onset of disease is correlating with age. Animal studies suggest that aging induces parkinsonian changes in dopaminergic neurons and accelerates the pathogenesis of PD (Collier, Kanaan, & Kordower, 2011; Schapira et al., 1990). A higher incidence of PD in men indicates that female hormones may be protective factors for PD.

Rural living, well water drinking, and farming were also evaluated as PD risk factors. However, these results were inconsistent because there were confounding factors such as herbicide and pesticide exposure (Wirdefeldt et al., 2011). Pesticide, herbicide, insecticide, and fungicide exposure is occupational PD risk (Priyadarshi, Khuder, Schaub, & Priyadarshi, 2001). Several epidemiological studies have suggested that heavy metal exposure including manganese, copper, and lead causes PD risk, particularly after a long exposure period (Coon et al., 2006; Elbaz & Moisan, 2008; Gorell et al., 1999; Wirdefeldt et al., 2011).

Various studies show that smoking decreases the risk of PD (Powers et al., 2008; Searles Nielsen et al., 2011; Tanaka et al., 2010; Wirdefeldt et al., 2011). A similar inverse correlation was found between coffee drinking and the risk of PD. Caffeine partially protects mice from 1-methyl-4-phenyl-1,2,3,6-tetrahydropyridine (MPTP)-induced neurodegeneration by altering striatal cytochrome P450 1A2 (CYP1A2), the adenosine A (2A) receptor, and dopamine transporter (DAT) expression (Singh et al., 2009). There was no substantial evidence indicating an effect of alcohol on PD development (Fukushima et al., 2010; Kiyohara et al., 2011).

Several studies have focused on the vascular risk factors in PD including diabetes, hypertension, and hypercholesterolemia. Only an association was found between high levels of total cholesterol and PD risk in a large Finnish prospective study (Hu, Antikainen, Jousilahti, Kivipelto, & Tuomilehto, 2008). Epidemiological studies do not support a role of obesity in PD pathogenesis (Chen et al., 2004; Palacios et al., 2011). Several studies found that exercise exerts neuroprotective effects in animal models of PD induced by 6-hydroxydopamine (6-OHDA) and MPTP (Lau, Patki, Das-Panja, Le, & Ahmad, 2011; Tajiri et al., 2010), and epidemiological research also showed that high physical activity is associated with a reduced risk for PD (Xu et al., 2010). Viruses or other organisms may be causative factors for parkinsonism. The parkinsonism that occurred after a viral encephalopathy following the 1918 influenza pandemic supports this hypothesis (Foley, 2009). Several viruses including influenza, Coxsackie, Japanese encephalitis B, and the human immunodeficiency virus (HIV) have been associated with both acute and chronic parkinsonism (Jang, Boltz, Webster, & Smeyne, 2009).

The effects of various drugs on PD risk have not been very well evaluated. Vitamins E and C are the most commonly investigated antioxidants in PD risk research studies. Recent case-controlled and meta-analysis studies suggested that only vitamin E exhibited a reduced risk for PD (Etminan, Gill, & Samii, 2005; Miyake et al., 2011).

2.3. Clinical findings

The main clinical features of PD include tremor, rigidity, and bradykinesia (Bartels & Leenders, 2009). Gait disturbance, flexion of the limb and trunk, dysarthria, hypophonia, and cognitive changes are other common clinical manifestations of PD that cause disability in patients. Nonmotor symptoms including impaired olfaction, sleep disturbances, visual hallucinations, constipation, orthostatic hypotension, and depression can occur very early in the disease course (Lim & Lang, 2010).

2.4. Treatment

Currently, there is no specific treatment for PD. However, current treatment methods do improve the quality of life in PD patients. Levodopa (L-DOPA) and peripheral dopa decarboxylase inhibitors are the initial treatment preferences in PD. L-DOPA treatment has motor side effects including dyskinesia and motor fluctuations (Lees, Hardy, & Revesz, 2009). Selegiline and rasagiline are selective monoamine oxidase type B inhibitors (MAO-B) that can be used in PD therapy; however, these agents are less effective in treating PD symptoms (Elmer & Bertoni, 2008; Lees et al., 2009). Amantadine, a dopamine agonist, is an initial choice in PD treatment. Recently, it has been demonstrated that amantadine diminishes the inflammatory activation of microglia (Kim et al., 2011). The catechol-O-methyl transferase inhibitors (COMTIs) are another pharmacological treatment alternative and are more effective in patients with *COMT* Val158Met polymorphism (Corvol et al., 2011). Alternative treatment strategies including deep brain stimulation (DBS), glial cell-derived neurotrophic factor (GDNF) treatment, adenosine A2 antagonism, and cell replacement therapies are still under investigation (Lees et al., 2009).

3. INFLAMMATION IN PD

3.1. Inflammation in neurodegenerative diseases

Neurodegenerative diseases are characterized by the dysfunction and loss of neurons and synapses in vulnerable areas of the nervous system with clinical outcomes such as memory loss, dementia, and movement disorders (Double, Reyes, Werry, & Halliday, 2010; Jellinger, 2009). Neurodegenerative diseases are classified into two groups based on onset and disease progression: acute and chronic. Acute neurodegenerative diseases include stroke and traumatic brain injury, while chronic neurodegenerative diseases include age-related and progressive diseases such as AD, in which tau and β-amyloid are deposited, PD, in which α-SYN is deposited, amyotrophic lateral sclerosis (ALS), Huntington's disease (HD), tauopathies, and age-related macular degeneration (Jellinger, 2009). The most prominent mechanisms of neurodegeneration involve abnormal protein accumulation as a result of protein misfolding, deficiency of protein degradation by a ubiquitin-proteasome system and the autophagy-lysosomal degradation pathway, formation of ROS and RNS, which causes oxidative stress, impaired bioenergetics and mitochondrial dysfunction, disruption of the neuronal Golgi apparatus and transport, endoplasmic reticulum stress, lack of neurotrophic factor support, and the progressive chronic neuroinflammatory process (Jellinger, 2009). All of the above-stated diseases proceed with neuronal death by several mechanisms, such as necrosis, apoptosis, autophagy, and necroptosis (Jellinger, 2009), although the pathological initiator is often divergent. During cell death, the brain's immune system is activated and, unless this activation is resolved, the neuroinflammatory environment in the brain is sustained and results in neuronal dysfunction (Tansey & Goldberg, 2010).

Inflammation, which is generated by an activated immune system, is a strictly regulated self-defensive mechanism against pathogenic stimuli or injury, which results in the protection of the host organism by clearance of pathogenic stimuli or debris to promote the healing process (Khandelwal, Herman, & Moussa, 2011; Stone, Reynolds, Mosley, & Gendelman, 2009). The immune system consists of two parts: innate and adaptive. The innate immune system, which consists of mononuclear phagocytes (monocyte, microglia, macrophage, and dendritic cells), natural killer cells, and neutrophils, is the first line of defense against pathogens and creates rapid, but short-term responses. On the other hand, the adaptive (acquired) immune system consists of T- and B lymphocytes and generates pathogen-specific, nonrapid, and long-lasting responses compared to the innate immune system (Stone et al., 2009). The genes that code for inflammatory molecules are suppressed under physiological conditions. Under stress conditions, which include infection or necrosis signals, danger or stranger nonself molecules are recognized by pattern recognition receptors (PRRs), such as Toll-like receptors (TLRs) and nucleotide oligomerization domain receptors (NLRs) (Amor, Puentes, Baker, & van der Valk, 2010; Glass et al., 2010; Hanke & Kielian, 2011). Stranger molecules, which are also known as pathogen-associated molecular patterns (PAMPs), are molecules that are specific to bacteria or virus, whereas danger molecules, which are also known as danger-associated molecular patterns (DAMPs), include heat-shock proteins, uric acid, chromatin, adenosine and ATP, high mobility group box chromosomal protein 1 (HMGB-1), β-amyloid, tau, and α-SYN (Lynch & Mills, 2012). The immune system in the CNS is an immune-privileged site, where microglia and astrocytes provide protection against external or internal pathogenic stimuli (Glass et al., 2010). Activation of the brain immune system is dependent on local TLRs and NLRs. Apart from the local activation of microglia, neuroinflammatory responses require cross talk between immune, vascular, and parenchymal cells (Glass et al., 2010). Furthermore, sustained and uncontrolled neuroinflammatory responses may result in the production of neurotoxic substances that induce neurodegenerative pathology (Khandelwal et al., 2011). In addition to its neurotoxic properties, microglia may also exert neuroprotective and regenerative effects (Gao & Hong, 2008). From this perspective, neurotrophic factor secretion by astrocytes and microglia promotes tissue repair and regrowth. Taken together, in neurodegenerative diseases, the beneficial effects of neuroinflammation are inadequate or ineffective and can result in the worsening of disease progression.

3.2. Hallmarks of inflammation

3.2.1 Microglia

Microglia are the key immune cells of the CNS, constituting up to 5–15% of the brain and 20% of the glial cell population (Kofler & Wiley, 2011; Smith, Das, Ray, & Banik, 2012). Microglia are originated from primitive macrophage progenitors in the yolk sac and migrate into the CNS during early embryogenesis (Saijo & Glass, 2011). Within healthy CNS tissue, microglia have a unique ramified morphology, consisting of a small and round soma with numerous branching processes (Smith et al., 2012). Microglia are active sensors that mediate innate immune responses in the brain via antigen-presenting and effector functions such as phagocytosis (Saijo & Glass, 2011). In addition to its immunological characteristics, microglia possess various beneficial functions, such as neurotrophic factor release, removal of toxic substances, neuronal repair, synaptic remodeling, synaptic pruning, and guidance to neural stem cells during synaptogenesis (Paolicelli et al., 2011; Ransohoff & Stevens, 2011; Saijo & Glass, 2011). Microglia are not uniformly dispersed in the brain parenchyma. In humans, microglia have been shown to localize mostly in the medulla oblongata, pons, basal ganglia, and substantia nigra (Mittelbronn et al., 2001). In addition to the parenchymal microglia, the CNS contains several other types of mononuclear phagocytes, namely, meningeal macrophages, choroid plexus macrophages, epiplexus cells, and perivascular macrophages (Ransohoff & Cardona, 2010). In comparison, the CNS mononuclear phagocytes express more major histocompatibility complex (MHC) class II molecules than microglia. Moreover, CNS mononuclear phagocytes interact with CD4+ memory T cells and effector T cells in the cerebrospinal fluid (CSF) and subpial vessels. In contrast, the parenchymal microglia stay behind the BBB to detect danger signals derived from a BBB breach (Ransohoff & Perry, 2009). Furthermore, the activation states and effector properties of microglia are affected by neurotransmitter release, such as γ-aminobutyric acid (GABA) (Lee, Schwab, & McGeer, 2011). Finally, at the molecular level, microglia express complement C1q subcomponent subunit A (C1qA), the triggering receptor expressed on myeloid cell 2 (TREM2), and the chemokine (C-X-C motif) ligand 14 (CXCL14) after activation stimuli, which are different than peripheral macrophages (Schmid et al., 2009).

In addition, microglia express many of the receptors that sense PAMPs and DAMPs, such as TLRs and NLRs (Saijo & Glass, 2011). TLRs are the primary surface receptors that recognize innate immune stimulants on microglia (Kofler & Wiley, 2011). Of the TLR receptors, TLR1–TLR9 are expressed in microglia (Parkhurst & Gan, 2010). Furthermore, NLRs, which are found in the cytosol, recognize cytosolic viruses and bacteria products. Upon ligand binding, the proinflammatory signaling pathways of both TLRs and NLRs are activated (Kofler & Wiley, 2011). Moreover, other members of the PRR family receptors, which do not have specific activator molecules, are also expressed on microglia, such as the scavenger receptors (CD36) and the C-type lectin domain receptors. Contrary to the PRR family receptors, microglia express most of the P2 purinoreceptors. The P2 purinoreceptors are classified into two categories: ionotropic, which are mainly activated by ATP binding, and metabotropic, which are activated by purines and pyrimidines, after which the signal is transduced into signaling pathways via G-proteins (Saijo & Glass, 2011).

Microglial activation occurs in response to various environmental challenges. In addition to activation by several bacterial and viral molecules, microglia can be stimulated by disease proteins (β-amyloid and α-SYN) and soluble mediators that are released by damaged neurons (Lull & Block, 2010). In the case of neuronal damage, μ-calpain, α-SYN, MMP-3, and neuromelanin are secreted from the injured neuron, which results in activation of microglia and subsequent microgliosis events (Lull & Block, 2010). During the activation process, transformation and proliferative events take place to form reactive microglia in different phenotypes (Czeh, Gressens, & Kaindl, 2011). In transformation, the resting "ramified" microglia phenotype shifts into the intermediate hyper-ramified morphology, which is characterized by a larger soma and an amoeboid morphology to initiate phagocytosis (Graeber & Streit, 2010). Aside from transformation and proliferation, microglia also upregulate cell surface markers of inflammation including MHC class I and II, cytokine receptors, and chemokine receptors (Jurgens & Johnson, 2012). Following activation in the early phases, microglia migrate through the site of damage and secrete pro- and anti-inflammatory cytokines [e.g., tumor necrosis factor-α (TNF-α), interleukin-1β (IL-1β), and interleukin-10 (IL-10)], chemokines, NADPH oxidase, prostaglandins, and oxidative and nitrosative stress-inducing factors, such as nitric oxide (NO) and superoxide ($O_2^{\bullet -}$) (Lull & Block, 2010). Microglial activation can also be acute or chronic depending on the type and duration of the external stimuli or activating factor (Czeh et al., 2011). Whereas short-term activation of microglia is generally believed to be neuroprotective, chronic activation has been implicated as a potential mechanism in neurodegenerative disorders. Thus, increased time of inflammation flares the amount of proinflammatory molecules and subsequently the neurons are eventually affected by neurodegenerative environment (Smith et al., 2012). As a consequence of the activation, microglia are divided into two subsets (M1 and M2) according to the distinct molecular phenotype and effector function, which are similar to the peripheral macrophage activation states (Czeh et al., 2011). In the M1 phenotype, which is also known as classical activation, the microglia produce a high amount of oxidative metabolites, proteases, and proinflammatory cytokines to defend the host organism against pathogens and tumor cells. However, in the M2 phenotype, which is also known as alternative activation, CD206 [alternatively known as mannose receptor c-type 1 (MRC1)] and arginase-1 are expressed by microglia. Here, tissue repair and remodeling are induced by the secreted anti-inflammatory cytokine IL-10 and transforming growth factor-β (TGF-β) (Czeh et al., 2011). The role of microglia in neurodegenerative diseases is unequivocal. However, the mechanism that underlies the switch from neuroprotective to autoaggressive effector microglia, which causes neurodegeneration, is yet unknown (Hanisch & Kettenmann, 2007; Khandelwal et al., 2011).

At the end of the inflammatory response, the inflammation must be resolved. However, if the resolution of inflammation is not completed, then the sustained inflammation causes neurotoxicity in the CNS, which can result in neurodegeneration. The resolution mechanism requires the termination of proinflammatory signaling pathways and the diminishment of inflammatory cells to restore normal tissue function (Serhan, 2011). The resolution of inflammation is tightly regulated by several posttranscriptional regulation and epigenetic mechanisms, including microRNAs (Alam & O'Neill, 2011; Rivest, 2009). Furthermore, during the neuroinflammatory response, secreted factors, such as Fas/Fas ligand and NO, also affect microglia, which causes cell death (Yang, Min, & Joe, 2007). Thus, this phenomenon is called activation-induced cell death (AICD). Moreover, some *in vitro* studies have demonstrated that cotreatment of microglia with interleukin-13 (IL-13) and lipopolysaccharide (LPS) causes upregulation of COX2 and production of PGE2, eventually leading to microglial cell death (Yang et al., 2006). Consequently, the inflammatory stimulators cause the death of the activated microglia by the indirect activation of apoptosis through the production of secondary mediators without any activation of the intrinsic apoptosis pathway.

The microglial activation process is affected in the aged individual and is characterized by a chronic low-level inflammation and increased reactivity. This phenotype is often referred to as "primed" or "sensitized" because of the more rapid induction of proinflammatory molecules compared to normal microglia (Streit & Xue, 2010). Primed or sensitized microglia are an intermediate activation state, which is evidenced by morphological activation (deramification) and upregulation of the MHC class II cell surface markers (von Bernhardi, Tichauer, & Eugenin, 2010). Apart from the MHC class II molecules, expression of the costimulatory molecules

CD80 and CD86 and the intercellular cell adhesion molecule-1 (ICAM-1) increases in the normal aged brain to acquire an activated immunophenotype (Downer et al., 2010; Frank, Barrientos, Watkins, & Maier, 2010). While the levels of proinflammatory cytokines elevate in the aged brain, the anti-inflammatory cytokines, IL-10 and interleukin-14 (IL-14), are downregulated (Dilger & Johnson, 2008). These findings are consistent with the results of gene expression profiling studies performed on aged human and rodent brains. Thus, aging is an important factor that influences the microglial activation process. Following a stimulus in the aged brain, the rapid and prolonged expression of proinflammatory molecules causes diseased behavior, cognitive impairment, and depressive behaviors (Streit, 2006).

In conclusion, microglia are resident innate immune cells of the CNS, which are stimulated by stranger (bacterial or viral molecules) and danger (cell debris) molecules. Upon activation, the microglia produce proinflammatory molecules to generate a neuroinflammatory response that mediates the removal of damaged cells by phagocytosis. Apart from its neurotoxic characteristics, microglia can also provide neurotrophic support for neurons by neurotrophic factor release, such as BDNF and GDNF. Furthermore, neuroinflammation can be detrimental and may eventually result in neurodegeneration unless the inflammatory response terminates in a timely fashion. In contrast, aging influences the dynamics of microglia by making them more susceptible to activation. However, most of these findings have been extrapolated from *in vitro* and *in vivo* animal studies and should be further studied in humans.

3.2.2 Astrocytes

Astrocytes, named after their morphology, are abundant glial cells in the CNS and are five times greater in number than neurons (Freeman, 2010). Astrocytes are dispersed throughout the CNS, where one astrocyte occupies its own territorial region without any overlap with the territory of another. They exhibit numerous branches that extend through neurons and blood vessels to form functional networks via gap junctions, which are called neurovascular units (NVUs) (Allaman, Belanger, & Magistretti, 2011). The mechanisms of astrocyte development and specification during embryonic development remain unknown (Sofroniew & Vinters, 2010). However, recent studies have elucidated the basic understanding of how astrocytes are produced by precursor cells and how astrocytes diversify by intrinsic and extrinsic factors (Freeman, 2010).

Astrocytes can be classified into two main groups, based on their location, processes, and morphology, namely, protoplasmic and fibrous astrocytes, which are found in the gray matter and white matter, respectively (Halliday & Stevens, 2011; Nimmerjahn, 2009). Protoplasmic astrocytes are mostly found around neuronal cell bodies, while fibrous astrocytes wrap around the nodes of Ranvier and oligodendrocytes (Oberheim et al., 2009; Sofroniew & Vinters, 2010). Finally, there are also morphological differences between rodent and primate astrocytes. According to a comparative study, protoplasmic astrocytes in the human neocortex are larger and contain more processes than rodent protoplasmic astrocytes (Oberheim et al., 2009).

In normal and pathological CNS tissue, astrocytes perform many essential functions. They have regulatory and supportive roles in the CNS, such as biochemical and nutritional support for neurons, extracellular ion balance, and repair of scarring of brain and spinal cord tissue (Sofroniew & Vinters, 2010). From a biochemical perspective, astrocytes metabolize and synthesize amino acids due to the expression of the enzyme pyruvate carboxylase, which is capable of replenishing intermediate molecules of metabolic reactions (Maragakis & Rothstein, 2006). Moreover, via specific shuttle systems, such as the malate–aspartate and glutamate–glutamine shuttle systems, astrocytes transport nutrients and metabolites to neurons. For example, cholesterol is required for normal brain function in the CNS. Cholesterol in the brain is synthesized by astrocytes and is utilized by neurons (Liu, Tang, et al., 2010). Outside nutritional support, astrocytes also maintain ion concentrations in the intracellular and extracellular spaces of the brain. To stabilize the pH, carbonic anhydrase converts carbon dioxide, which is produced by neurons as a result of the oxidative metabolism of pyruvate, to hydrogen and bicarbonate ions (Maragakis & Rothstein, 2006).

Moreover, astrocytes form a bridge between the vascular system and neurons, which makes them a key regulatory element of neuronal activity and cerebral blood flow. Potassium ions may be retained following neuronal activity, and astrocytes provide homeostasis via the expression of potassium channels around synapses and capillaries (Kofuji & Newman, 2004). In addition, during synaptic transmission, the neurotransmitter glutamate exhibits excitotoxic effects, which may be released into the synapse and subsequently activates metabotropic glutamate receptors on astrocytes. The activation of metabotropic glutamate receptors triggers the release of arachidonic acid metabolites, which causes Ca^{2+} release as a result of activating the inositol triphosphate (IP_3) pathway at the astrocytic endfeet. If this activation occurs near a blood vessel, it results in the dilation of arterioles (Takano et al., 2006).

Furthermore, astrocytes may regulate oxidative stress to maintain CNS homeostasis. For this reason, nuclear factor E2-related factor 2 (Nrf2), which is an antioxidant transcription factor expressed in astrocytes, is activated to induce the expression of the antioxidant response element (ARE), which bears phase II detoxification and cytoprotective genes, such as hemeoxygenase-1 and glutathione S-transferase (Calkins et al., 2009; Vargas, Johnson, Sirkis, Messing, & Johnson, 2008). Additionally, Nrf2 activation causes a reduction of ROS production following interferon-γ (IFN-γ) stimulation, demonstrating that astrocytes are involved in antioxidant signaling (Chung et al., 2010).

Astrocytes also have immune functions that limit the spreading of inflammatory cells and infectious agents from the damaged area to healthy parenchyma (Hamby & Sofroniew, 2010). From this perspective, the astrocytes play both beneficial and harmful roles. In the case of astroglial activation, the cells undergo reactive astrogliosis, which includes scar formation. In this process, the different isoforms of glial fibrillary acidic protein (GFAP), which is an intermediate filament protein expressed in astrocytes, are upregulated in mature astrocytes (Middeldorp & Hol, 2011). During reactive astrogliosis, astrocytes undergo morphological changes that result in cell hypertrophy without any effect on cellular domains and tissue structure (Hamby & Sofroniew, 2010; Sofroniew & Vinters, 2010). Furthermore, during the activation process, astrocytes act as immune regulatory cells by antigen presentation and their expression of pro- and anti-inflammatory molecules, such as cytokines, chemokines, complement factors, and neurotrophic factors (Allaman et al., 2011; Ricci, Volpi, Pasquali, Petrozzi, & Siciliano, 2009; Sidoryk-Wegrzynowicz, Wegrzynowicz, Lee, Bowman, & Aschner, 2011; Sofroniew & Vinters, 2010). Moreover, immediately after activation, astrocytes further activate distant microglia through a calcium wave within the astrocytic network after induction of IP_3 and thereby increase the amount of cytosolic Ca^{2+} (Liu, Tang, & Feng, 2011).

Most functions of the brain that are mostly regulated by astrocytes are affected by aging. The most observed alteration in aging is the overexpression of GFAP in hippocampal formation, which results in the flattening of astroglial morphology (Salminen et al., 2011). As organisms age, the BBB, which is composed of NVU by astrocytes, deteriorates, leading to a cognitive decline and dementia (Zlokovic, 2008). The disruption of the BBB is mainly caused by the age-related disturbance of Ca^{2+} signaling and secretion of interleukin-6 (IL-6), TNF-α, and IL-1β by astrocytes (Salminen et al., 2011).

In conclusion, astrocytes are the most prominent cells of the CNS, and they have many essential functions that regulate environmental homeostasis and neuronal activity. They also mediate inflammatory responses in the brain by secreting pro- and anti-inflammatory molecules. During neurodegenerative pathology, astrocytes also contribute to sustained inflammation; therefore, they are also responsible for neurodegenerative diseases.

3.2.3 Adaptive immune system

The adaptive immune system is composed of T- and B lymphocytes, which have effector and memory functions (Sriram, 2011). T cells have roles in cell-mediated immune responses. There are several subsets of T cells, namely, T_H (helper T cells), T_C (cytotoxic T cells), memory T cells, T_{reg} (regulatory T cells), NKT (natural killer T cells), and $\gamma\delta$ T cells (Cao, Li, & Shen, 2011). T_H cells are CD4+ T cells, which have several subtypes such as T_{H1}, T_{H2}, and T_{H17}, which regulate or assist in immune responses by their cytokine expression. T_C cells express CD8 on their cell surfaces and have cytotoxic effects that destroy virus-infected or tumor cells. Memory T cells persist at the peripheries after long periods of initial exposure to antigen to prevent the repeat of infection. Finally, T_{reg} cells are required for the maintenance of immunological tolerance by shutting down T cell-mediated immunity and suppressing autoreactive T cells that have escaped negative selection during development (Gendelman & Appel, 2011). Naïve T cells are activated by antigen-presenting cells (APCs) via T cell receptors and several other surface molecules as well as soluble mediators that are secreted by innate immune system cells, such as cytokines and chemokines. Following activation, T cells produce cytokines that have an autocrine effect to facilitate their own maturation process, where the activated T cells behave according to their subtype. During the activation process, the T_{reg}/T_{eff} (effector T cell) balance determines the result of the inflammatory response (Ha, Stone, Mosley, & Gendelman, 2012). In the CNS, infiltration of T lymphocytes was demonstrated in neurodegenerative diseases (Rezai-Zadeh, Gate, & Town, 2009). In addition, B cells mediate humoral immune responses, which are driven by secreted soluble antibodies. B cells can also be costimulated by APCs and T_H cells. Soluble antibodies travel around the body within the bloodstream, find the antigen, and bind to mark them for destruction. After labeling cells for further destruction, the constant Fc domain, which contains immunoglobulin (Ig), binds to the antibody to form the recognition complex. This complex is recognized by a cell surface receptor (FcR) for destruction by cytokine production, phagocytosis, and degranulation (Okun, Mattson, & Arumugam, 2010). FcRs are expressed by lymphocytes, macrophages, and mast cells in the periphery. Moreover, various FcRs are expressed by neurons, astrocytes, oligodendrocytes, and microglia (Okun et al., 2010). Finally in the CNS, two types of FcRs have been found to be associated with the pathogenesis of AD (Deane, Bell, Sagare, & Zlokovic, 2009) and PD (Orr, Rowe, Mizuno, Mori, & Halliday, 2005).

3.2.4 Soluble protein mediators

During the inflammatory response, a number of mediator molecules are secreted into the environment by immune cells, such as cytokines and chemokines. Cytokines and chemokines are messenger proteins that mediate communication between immune system cells and other cells (Fuxe, Tarakanov, Goncharova, & Agnati, 2008; Helmy, De Simoni, Guilfoyle, Carpenter, & Hutchinson, 2011). Within these cells, cytokines regulate immune responses and coordinate immune cell interactions. Furthermore, chemokines are defined as chemotactic cytokines (Rostene et al., 2011). Cytokines and chemokines generally have pro- or anti-inflammatory effects. These molecules are also known to associate with physiological processes, such as learning, memory, neural plasticity, and neurogenesis (Okun, Griffioen, & Mattson, 2011; Spooren et al., 2011; Yirmiya & Goshen, 2011).

During inflammatory responses, the complement system cascade exhibits scavenger roles between the innate and adaptive immune systems. In the complement cascade, cytolytic membrane attack complexes (MAC) and opsonin molecules target pathogenic and toxic molecules to identify "nonself" molecules (Griffiths, Gasque, & Neal, 2010). Furthermore, at the end of the inflammatory response, apoptotic cells and debris are removed from the site of inflammation by the complement system. Three pathways, namely, the classic complement pathway, alternative complement pathway, and mannose-binding lectin pathway, regulate the activation of the complement system (Horstman et al., 2012; Veerhuis, Nielsen, & Tenner, 2011). In the CNS, the balance between complement activators and complement regulatory proteins are disturbed, which results in the occurrence of neurodegenerative pathology (Horstman et al., 2012; Veerhuis et al., 2011).

4. EVIDENCE OF INFLAMMATION IN PD

4.1. Postmortem studies

The first evidence for a role of neuroinflammation in PD originated from a study conducted by McGeer, Itagaki, Boyes, and McGeer (1988). They showed large numbers of human leukocyte antigen DR (HLA-DR)-positive reactive microglia in the SN of patients with PD. This initial finding was later confirmed by other postmortem studies (Desai Bradaric, Patel, Schneider, Carvey, & Hendey, 2012; Imamura et al., 2003; Mirza, Hadberg, Thomsen, & Moos, 2000). Reactive astrocytosis was determined using immunohistochemistry with antibodies against GFAP and glutathione peroxidase (Gpx). Increased density of Gpx-positive cells was found around the dopaminergic neurons in postmortem tissues (Damier, Hirsch, Zhang, Agid, & Javoy-Agid, 1993). In a recent study, GFAP-positive cells were mainly present in the caudate nucleus and rare in the putamen. The immunostaining showed variability between patients, and there was no correlation between GFAP staining and disease severity (Mythri et al., 2011). T lymphocytes (CD8+ and CD4+ T cells) were also identified in postmortem human brain specimens of PD patients (Brochard et al., 2009). Endothelial cells may also contribute to disease pathogenesis. In addition, an increased number of endothelial cell nuclei in the SNpc of patients were found (Faucheux, Bonnet, Agid, & Hirsch, 1999). Moreover, ultrastructural abnormalities of cerebral microvessels, including capillary basement membrane thickening and collagen accumulation in the basement membrane, were determined in the postmortem brains of PD patients (Farkas, De Jong, de Vos, Jansen Steur, & Luiten, 2000). Furthermore, $\alpha v\beta 3$ integrins may act as an angiogenic marker that is elevated in the locus coeruleus (LC), SNpc, and putamen in PD patients (Desai Bradaric et al., 2012).

Evidence of inflammation in PD has been evaluated in postmortem brain samples using various molecular biological methods. Several cytokine levels, including TNF-α, TNF-α receptor, IL-1β, IL-6, IFN-γ, and the inflammation-related transcription factor NFκB, were elevated in the SN (Boka et al., 1994; Mogi, Harada, Kondo, et al., 1994; Mogi, Harada, Riederer, et al., 1994; Mogi, Kondo, Mizuno, & Nagatsu, 2007). Furthermore, TLR9, an inflammation-related receptor, was upregulated in the striatum of PD patients (Ros-Bernal et al., 2011). In addition, iNOS and COX-1 and -2 were found to increase in the parkinsonian brain (Knott, Stern, & Wilkin, 2000). Microarray studies also confirmed the upregulation of inflammation-related genes including *NFκB*, HLA-C, and cytotoxic T lymphocyte antigen 4 (*CTLA-4*) in the SN of postmortem brains (Moran & Graeber, 2008; Moran et al., 2007; Simunovic et al., 2009). Proteomic analysis of human SN in PD verified the upregulation of GFAP expression (Basso et al., 2004). Cell-type-specific mRNA or protein expression changes could also be determined in individual laser capture microdissected cells. This approach was used for the analysis of mRNA expression changes in SN dopaminergic neurons of PD patients (Grundemann, Schlaudraff, Haeckel, & Liss, 2008).

4.2. Imaging studies

In vivo imaging of microglial activation provides evidence of inflammation in PD. [11C](*R*)-PK11195, which binds to the mitochondrial membrane translocator protein-18 in the benzodiazepine binding side complex, is used for tracking microglial activation. The levels of [11C](*R*)-PK11195 in the midbrain are significantly higher in PD than in controls and are correlated with the motor severity of the disease (Ouchi et al., 2005). However, Gerhard et al. found that microglial activation was not correlated with disease severity or duration (Gerhard et al., 2006). In fact, microglial activation in PD was not altered during 2 years of follow-up. These studies indicate the difficulty associated with using this compound; thus, a new PET tracer is still necessary for the determination of microglial activation (Stoessl, Martin, McKeown, & Sossi, 2011). [(11)C]-verapamil positron emission tomography (PET) was also used to investigate the function of the BBB P-glycoprotein (P-gp). The BBB-P-gp function was not altered in PD patients (Bartels et al., 2008).

4.3. Analysis of CSF samples

Inflammatory markers were analyzed in the CSF of PD patients. The levels of TNF-α, IL-1β, and IL-6 were elevated (Blum-Degen et al., 1995; Mogi, Harada, Kondo, et al., 1994; Mogi, Harada, Riederer, et al., 1994; Zhang et al., 2008), whereas the TGF-β1, interleukin-12 (IL-12), IFN-γ, and IL-10 levels did not change in the CSF of PD patients (Rota et al., 2006). In addition, a significant inverse correlation between disease severity and IL-6 CSF levels was found (Muller, Blum-Degen, Przuntek, & Kuhn, 1998). The levels of the complement system components were altered in the CSF of patients with PD (Finehout, Franck, & Lee, 2005). Several proteomic studies have investigated the altered proteins in the CSF of PD patients (Constantinescu et al., 2010; Sinha et al., 2009; Yin et al., 2009). In one proteomic study, CD14, a glial function regulator, was found at high levels in the CSF (Yin et al., 2009). Furthermore, PD-CSF reduces microglial and astrocyte growth and increases α-SYN in microglial cells (Schiess et al., 2010). These results support a role of inflammation in PD.

4.4. Peripheral immune evidence

Peripheral evidence supports a role of inflammation in PD pathogenesis. Levels of interleukins [interleukin-2 (IL-2), interleukin-4 (IL-4), IL-6, IL-10], macrophage migration inhibitory factor, TNF-α, TNFαR1, INFγ, RANTES, and osteopontin were elevated in the serum of PD patients (Brodacki et al., 2008; Dufek et al., 2009; Hofmann et al., 2009; Maetzler et al., 2007; Nicoletti et al., 2011; Rentzos et al., 2007, 2009; Scalzo, Kummer, Cardoso, & Teixeira, 2009, 2010). The proportion of peripheral immune cells was also altered in PD patients. The number of CD4+ lymphocytes was diminished, and the number of CD8+ lymphocytes was elevated in patients with PD (Baba, Kuroiwa, Uitti, Wszolek, & Yamada, 2005). Another study found that helper T cells and B cells were decreased and CD4+ CD25+ lymphocytes were increased in PD patients (Bas et al., 2001). In addition, elevated Fas expression was found in circulating CD4(+) helper T cells (Calopa, Bas, Callen, & Mestre, 2010). A subgroup of T cells (γδ+) was also found to be elevated in patients with PD (Fiszer et al., 1994). Immune cell alternations in PD support the notion of immune pathogenesis of PD.

4.5. Genetic risk factors in PD

Several genetic studies have investigated the association between polymorphism in immune-related genes and PD. Polymorphisms in HLA genes are the most evaluated risk factor for PD (Hamza et al., 2010; Lampe et al., 2003; Liu, Cheng, et al., 2011; Nalls et al., 2011; Puschmann et al., 2011; Saiki et al., 2010; Simon-Sanchez et al., 2011). A recent meta-analysis of GWAS confirmed an association between HLA-DRB5 and PD (Nalls et al., 2011). SNPs in the promoter region of the *TNF-α* gene (-G308A, -T1031C) also increase the risk of PD (Bialecka et al., 2008; Kruger et al., 2000; Wahner, Bronstein, Bordelon, & Ritz, 2007). The G308A polymorphism leads to an overexpression of TNF-α (Wilson, Symons, McDowell, McDevitt, & Duff, 1997). Many other inflammation-related gene polymorphisms, including CD14 monocyte receptor, interleukin-1α (*IL-1α*), *IL-1β*, *IL-6*, interleukin-8 (*IL-8*), and interleukin-18 (*IL-18*), are associated with PD risk (Hakansson et al., 2005; Lin, Chen, Yueh, Chang, & Lin, 2006; Mattila et al., 2002; McGeer, Yasojima, & McGeer, 2002; Nishimura, Kuno, Kaji, Yasuno, & Kawakami, 2005; Nishimura et al., 2001; Ross et al., 2004; Schulte et al., 2002; Xu et al., 2011; Zhou, Yang, Zhang, Wang, & Chan, 2008). Although inflammation-related gene polymorphisms support a role of inflammation in PD, large-scale GWAS studies are necessary to determine the exact risk of these SNPs in PD pathogenesis.

4.6. Epidemiological studies (NSAIDs, steroids)

The results of epidemiological studies to determine PD risk in the nonsteroidal anti-inflammatory drug (NSAID) user population support a role of inflammation in PD pathogenesis (Bornebroek et al., 2007; Bower, Maraganore, Peterson, Ahlskog, & Rocca, 2006; Chen et al., 2003, 2005; Gagne & Power, 2010; Gao, Zhang, et al., 2011; Hancock et al., 2007; Hernan, Logroscino, & Garcia Rodriguez, 2006; Samii, Etminan, Wiens, & Jafari, 2009; Ton et al., 2006; Wahner et al., 2007). An early study reported that two or more tablets of aspirin per day diminish PD risk (Chen et al., 2003). However, the same group was unable to replicate their findings (Chen et al., 2005); they did not find any association between aspirin intake and PD in a second study. Rather, they reported that only ibuprofen decreases the risk of PD (Chen et al., 2005). A large-scale study and a meta-analysis also revealed conflicting results regarding NSAIDs, with the exception that ibuprofen provides protection against PD (Gao, Zhang, et al., 2011; Samii et al., 2009).

5. INFLAMMATION IN ANIMAL MODELS OF PD

5.1. Toxin-based models

5.1.1 6-Hydroxydopamine

6-OHDA is a selective catecholaminergic neurotoxin that is widely used in animal models of PD (Bove & Perier, 2012). Injection of 6-OHDA into the striatum causes retrograde degeneration of nigrostriatal neurons (Salama & Arias-Carrion, 2011). Because 6-OHDA cannot cross the BBB, it has to be injected into the brain using stereotaxic procedures (Jackson-Lewis et al., 2012). The medial forebrain bundle, the SNpc, or the striatum can be targeted for 6-OHDA injection (Fulceri et al., 2006). Bilateral 6-OHDA administration leads to a severe disease phenotype; therefore, a unilateral lesion is preferred. Affected rats show rotational behavior as a response to dopaminergic agonists, such as apomorphine, due to striatal imbalance (Willis & Kennedy, 2004). The neurotoxic effect of 6-OHDA is due to the oxidative stress triggered by ROS production. This model does not recapitulate the pathological findings of PD, such as LBs (Jackson-Lewis et al., 2012).

Cellular and molecular evidence of inflammation are observed in the 6-OHDA-induced animal model of PD. Reactive astrocytosis was demonstrated with GFAP immunohistochemistry (Gomide, Bibancos, & Chadi, 2005; Wachter et al., 2010). Microglial activation also occurs in this model (Akiyama & McGeer, 1989; Marinova-Mutafchieva et al., 2009). 6-OHDA treatment does not increase the percentage of activated lymphocytes in rat PBMCs (Bas et al., 2001). The inflammatory mediator TNF-α was elevated in SN and striatum of a 6-OHDA-induced animal model of PD. CX3CL1 and CD200-CD200R signaling and TNF inhibition have neuroprotective effects in the 6-OHDA model (Harms et al., 2011; Pabon, Bachstetter, Hudson, Gemma, & Bickford, 2011; Zhang, Wang, et al., 2011). In addition, a proteomic analysis showed that an inflammation-related protein (complement component 1q subcomponent-binding protein) was altered in 6-OHDA-induced PD striatum (Park et al., 2010).

5.1.2 1-Methyl-4-phenyl-1,2,3,6-tetrahydropyridine

MPTP is a potent neurotoxin used to mimic PD in a wide range of organisms, including nonhuman primates, guinea pigs, mice, and cats (Chiueh et al., 1984; Meredith, Totterdell, Potashkin, & Surmeier, 2008; Schneider, Yuwiler, & Markham, 1986). MPTP rapidly crosses the BBB and is converted into the toxic metabolite MPP+ (Johannessen, 1991). Several MPTP administration methods have been used to develop acute, subacute, and chronic models. Daily i.p. injections of 30 mg/kg MPTP for five consecutive days is used in subacute model and causes a 30–40% dopaminergic cell loss in SN (Drolet, Behrouz, Lookingland, & Goudreau, 2004). MPTP can be administered systemically or via intracarotid infusion (Bankiewicz et al., 1986; Langston, Forno, Rebert, & Irwin, 1984). Another group used the intranasal pathway to administer MPTP (Prediger et al., 2006). There are several parameters such as animal strain, gender, age, and body weight that affect MPTP toxicity in mice (Bove & Perier, 2012). MPTP causes transient PD in young mice, but it leads to permanent nigrostriatal neuronal loss in aged mice (Gupta et al., 1986). Furthermore, female mice show higher mortality rates than males (Antzoulatos, Jakowec, Petzinger, & Wood, 2010). MPP+ produces neurodegeneration through the blockade of the electron transport chain enzyme, including complexes I, III, and IV (Desai, Feuers, Hart, & Ali, 1996). Many other factors contribute to MPTP toxicity, including iron, vesicular monoamine transporter (VMAT2), ROS, and apoptosis (Salama & Arias-Carrion, 2011).

Inflammation markers were also evaluated in the MPTP model. HLA-DR-positive reactive microglia have been found in the SN of monkeys following MPTP administration (McGeer, Schwab, Parent, & Doudet, 2003). Further, T lymphocyte infiltration was observed in the brain (Brochard et al., 2009). CD8+ T cells were more abundant than CD4+ T cells in the SN of mice with parkinsonism (Brochard et al., 2009).

There are strain differences for MPTP toxicity in mice. Microglial and astroglial activation and increased cytokine levels in CSF [IL-10, IL-12(p40), IL-13, IFN-γ, MCP-1, and TNF-α] were observed in MPTP-sensitive B6 but not in BALB/C mice (Yasuda et al., 2008). Similarly, IFN-γ and TNF-α levels in serum and CNS were increased in chronic parkinsonian macaques, and IFN-γR signaling was observed in increased numbers of activated glial cells in the SNpc (Barcia et al., 2011). Increased expression of CXCR4 and CXCL12 was found in the SN of MPTP-treated mice (Shimoji, Pagan, Healton, & Mocchetti, 2009). Aging may affect microglial responses. Microglial activation in an MPTP model of PD was more severe in aged SAMP8 mice (Liu, Wang, et al., 2010). A microarray analysis confirmed increased expression of inflammation-related genes, including cytokines [IL-1β, IL-6, interleukin-7 (IL-7), IL-10], cytokine receptors (IL-1R, IL-3R, IL-4R, IL-10R), and inflammation-related transcription factor NFκB in SN and striatum (Grunblatt, Mandel, Maor, & Youdim, 2001). Another microarray analysis in monkey determined that inflammation-related gene expression such as interleukin-11 (IL-11), chemokines, and complement system genes were elevated in SN and striatum (Ohnuki, Nakamura, Okuyama, & Nakamura, 2010).

5.1.3 Paraquat

1,1-Dimethyl-4,4-bipyridium (paraquat, PQ) is an herbicide that has been used as a possible parkinsonism-inducing agent (Berry, La Vecchia, & Nicotera, 2010). PQ has a similar structure to MPTP, causes degeneration of dopaminergic nigrostriatal neurons, and induces α-SYN-containing inclusion bodies (Manning-Bog et al., 2002). The lack of reproducibility with PQ may be due to age and strain differences; aged mice are more sensitive to PQ toxicity (Prasad et al., 2009). Inflammation and apoptosis mediate PQ toxicity (Bove & Perier, 2012).

PQ induces inflammation responses in an animal model of PD. Microglial activation was found in the SN of rats that received PQ with or without maneb (Cicchetti et al., 2005; Saint-Pierre, Tremblay, Sik, Gross, & Cicchetti, 2006; Yadav, Gupta, Srivastava, Srivastava, & Singh, 2011).

Even a single dose of PQ induces a microglial response without neuronal degeneration (Purisai et al., 2007). The expression of inflammation-related proteins including TNF-α, IL-1β, and NFκB was increased in the SN of PQ-induced animals (Mitra, Chakrabarti, & Bhattacharyya, 2011; Yadav et al., 2011). Concomitant with these findings was the observation that PQ causes less inflammatory response in IFN-γ knockout (KO) mice (Mangano et al., 2011).

5.1.4 Rotenone

Rotenone is a pesticide and mitochondrial complex I inhibitor (Greenamyre, Cannon, Drolet, & Mastroberardino, 2010). Chronic exposure to rotenone results in dopaminergic neuron degeneration and cytoplasmic inclusion formation (Jackson-Lewis et al., 2012). Rotenone can be administered by stereotaxic injection and systemic administration. Chronic rotenone administration may be more useful due to inconsistent and unpredictable effects of this agent when used acutely (Bove & Perier, 2012).

Microglial activation is present in a rotenone-induced animal model of PD, and minimal astrocytosis accompanies microglial activation (Sherer, Betarbet, Kim, & Greenamyre, 2003). Glial cell activation is more severe in older rats (Phinney et al., 2006). Apart from the SN, unilateral rotenone infusion into the medial forebrain bundle leads to a generalized increase of astrocytes and microglia (Norazit, Meedeniya, Nguyen, & Mackay-Sim, 2010).

5.2. Genetic models

5.2.1 Evidence of inflammation in transgenic animal models of PD

Several transgenic animal models of PD have been developed that express genes associated with human disease. The first α-SYN transgenic mouse was developed by Masliah et al. (2000). A platelet-derived growth factor subunit B (PDGFB) promoter was used to express human α-SYN in neuronal cells (Masliah et al., 2000). Evidence of inflammation was observed in this transgenic model. Astroglial activation was also observed in transgenic mice harboring a transgene with the Thy1 promoter in front of the *SNCA* gene (Rockenstein et al., 2002), and microglial activation and high levels of TNF-α were found in the SN of mice expressing human α-SYN under control of the rat TH promoter (Su et al., 2008). In another model, expression of truncated human α-SYN increases the number of CD11b-positive microglia in the SN of transgenic mice (Tofaris et al., 2006). Furthermore, transgenic mice that contain the most common human gene for mutated α-SYN (A53T α-SYN) show astrogliosis in the spinal cord (Giasson et al., 2002). Overexpression of human α-SYN in rat midbrain by using a recombinant viral vector increases microglia cell numbers without neurodegeneration (Sanchez-Guajardo, Febbraro, Kirik, & Romero-Ramos, 2010). Later on, those animals have neuronal death and increased numbers of CD68-expressing microglial cells and CD4+ and CD8+ T cells in the SN (Sanchez-Guajardo et al., 2010). Furthermore, overexpression of α-SYN elevates the levels of proinflammatory cytokines, including ICAM-1, IL-1α, IL-6, and TNF-α (Theodore, Cao, McLean, & Standaert, 2008). α-SYN increases the susceptibility of dopaminergic neurons to a subsequently administered toxin. For instance, LPS activates microglial cells, but it only causes neuronal death in transgenic mice that express normal or mutant (A53T) α-SYN (Gao et al., 2008). Further persistent neuroinflammation and neurodegeneration of the nigrostriatal dopamine pathway were seen only in *SNCA* transgenic mice (Gao, Chen, Schwarzschild, & Ascherio, 2011).

Several transgenic models were produced that contain mutant or knockout PD-related genes, including *parkin*, *PINK*, *LRRK2*, and *DJ-1*. There is no obvious evidence of inflammation in these models (Schwab, Klegeris, & McGeer, 2010). Increased proportions of astroglia and microglia were only seen in parkin-deficient mice after rotenone induction (Casarejos et al., 2006).

5.2.2 Relationship between inflammation and PD in inflammation-related gene studies

Transgenic mice deficient in inflammation-related genes were used to determine if specific molecules are associated with the inflammation pathogenesis of PD. Deficiency in cytokines (IL-1α/β, IL-6, IL-18, TNFRs), cell signaling molecules [MyD88, Smad 3, MAPK-activated protein kinase 2 (MK2), caspases, protease-activated receptor 1 (PAR1), MCP-1/CC chemokine ligand 2 (Mcp-1/Ccl2), prostaglandin E2 receptor subtype 2 (EP2)], and many other inflammation-related genes alter susceptibility to toxins that induce PD (Bolin, Strycharska-Orczyk, Murray, Langston, & Di Monte, 2002; Cote, Drouin-Ouellet, Cicchetti, & Soulet, 2011; Furuya et al., 2004; Hamill et al., 2007; Jin et al., 2007; Pattarini, Smeyne, & Morgan, 2007; Qin et al., 2007; Sugama et al., 2004; Tapia-Gonzalez et al., 2011; Thomas et al., 2008; Vroon et al., 2007). These findings strongly suggest that inflammation-related molecules participate in PD pathogenesis.

5.3. Inflammatory models

5.3.1 Lipopolysaccharide

LPS, an endotoxin found in the outer membrane of Gram-negative bacteria, induces a strong immune response in animals (Liu & Bing, 2011). In 1998, an animal model was reported in which a single intranigral LPS injection produces PD (Castano, Herrera, Cano, & Machado, 1998). Later, several administration routes were used for this PD model, including intrapallidal, intraperitoneal, and systemic (Qin et al., 2007; Wu et al., 2011; Zhang, Stanton, et al., 2005). Several experimental considerations affect this model, including the administration route and dose of LPS, strain, gender, and age of experimental animals (Tufekci, Genc, & Genc, 2011). The LPS model provides evidence of an inflammatory component in PD pathogenesis. Several observations substantiate this evidence, namely: (i) LPS activates astrocyte and microglia in the SN of animals with PD even after systemic administration (Iravani et al., 2005; Qin et al., 2007); (ii) LPS releases a wide variety of proinflammatory factors including IL-1α, TNF-α, IL-1β, and iNOS (Hernandez-Romero et al., 2008; Tomas-Camardiel et al., 2004; Zhou et al., 2005); and (iii) LPS also increases COX2 expression in the SN of PD animals (de Meira Santos Lima et al., 2006; Geng et al., 2011), whereas ROS and MMP-3 are other mediators of LPS-induced neurodegeneration (McClain, Phillips, & Fillmore, 2009; Qin et al., 2004).

5.3.2 Polyinosinic:polycytidylic acid

Polyinosinic:polycytidylic acid [poly(I:C)] is a TLR3 agonist molecule that was used in a rat model of PD (Deleidi, Hallett, Koprich, Chung, & Isacson, 2010). It was administered to the SN at a dose of 10–40 μg with microinfusion pumps to induce PD. Single poly(I:C) administration does not cause SN neurodegeneration, but it increases the susceptibility of dopaminergic neurons to a subsequent low dose of 6-OHDA (Deleidi et al., 2010). In addition, it causes a long-lasting inflammatory reaction in the SN and dorsolateral striatum that involves microglia, astrocytes, and perivascular and parenchymal CD68+ macrophages. The authors also found that the chemokines, MCP-1, and RANTES were significantly elevated in the SN, and IL-1β, IL-6, TNF-α, MCP-1, and TGF-β1 were elevated in the dorsolateral striatum of poly(I:C)-induced rats. IL-1R antagonist treatment rescued neurons from poly(I:C) and 6-OHDA toxicity (Deleidi et al., 2010).

5.3.3 *Prostaglandin J2*

Prostaglandin D2 is the major prostaglandin in the mammalian brain that produces prostaglandin J2 (PGJ2). It is a very toxic molecule that disturbs the UPS pathway (Li et al., 2004). Recently, Pierre et al. used it to develop another inflammation-related animal model of PD (Pierre, Lemmens, & Figueiredo-Pereira, 2009). PGJ2 administration to the SN causes neurodegeneration and α-SYN-containing protein aggregates. In addition, 6.7- and 16.7-μg doses of PGJ2 activate microglia and astrocytes (Pierre et al., 2009).

6. MOLECULAR MECHANISMS OF INFLAMMATION IN PD

In PD pathology, inflammation participates in disease progression (Fig. 4.1). In this process, microglia have primary effector immune functions by synthesizing toxic molecules. In addition to microglia, astrocytes are the secondary effector immune cells of the brain and support proinflammatory functions of microglia (Hirsch & Hunot, 2009). In PD models, dopaminergic neurons release soluble neuron injury factors, namely, MMP-3 (Kim et al., 2007), α-SYN (Zhang, Wang, et al., 2005), neuromelanin (Zecca, Zucca, Wilms, & Sulzer, 2003), and μ-calpain (Levesque et al., 2010; Lull & Block, 2010), which in turn activate microglia.

Extracellular α-SYN deposition is a pathological hallmark of PD (Lee, 2008). However, according to a very recent study, the idea that α-SYN is a cytosolic protein is refuted. This study suggests that α-SYN is physiologically secreted by neurons *in vivo* (Emmanouilidou et al., 2011). The secreted or deposited α-SYN can activate both microglia (Beraud et al., 2011) and astrocytes (Lee, Kim, & Lee, 2010) via TLR signaling. Furthermore, nitrated α-SYN is found to interact with CD4+ and CD4+ CD25+ T cells and then alters the microglial proteome to promote an activated state (Reynolds, Stone, Mosley, & Gendelman, 2009a, 2009b), which mediates proinflammatory responses and phagocytosis. Moreover, α-SYN can induce MMP-3 expression in microglia (Lee, Woo, et al., 2010). MMP-3 upregulation can be accompanied by the production of cytotoxic molecules and proinflammatory cytokines that promote neuronal damage (Kim & Hwang, 2011). Besides, neuromelanin was shown to activate microglial cells *in vitro*, resulting in secretion of oxidative stress-inducing molecules and proinflammatory cytokines (Zhang, Phillips, et al., 2011). When neuromelanin is injected into the SN, it causes rapid microglial activation followed by neurodegeneration (Zecca et al., 2008).

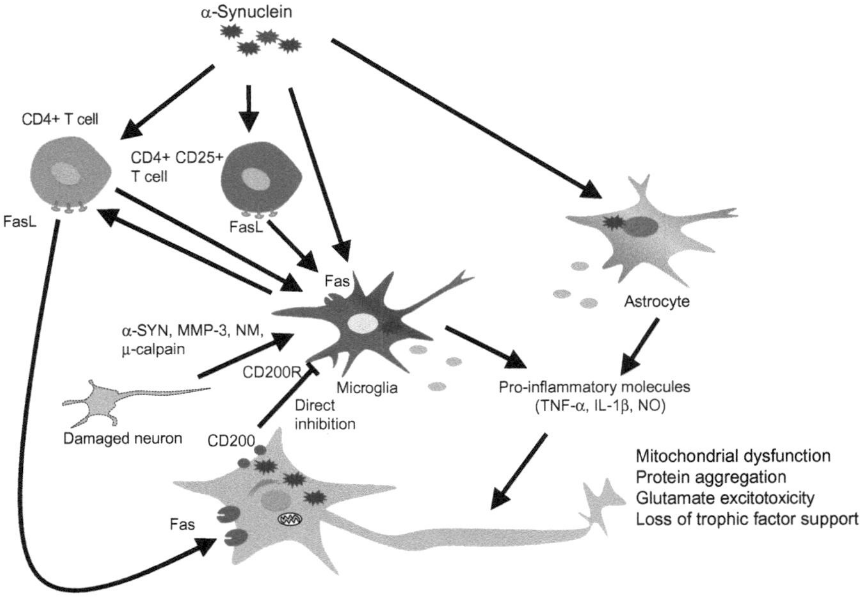

Figure 4.1 In PD pathogenesis, α-synuclein and stimulating factors released from damaged neurons trigger activation of microglia and astrocytes, which release proinflammatory molecules. Furthermore, activated microglia may secrete signals to recruit CD4+ T cells, which directly affect neurons via Fas/Fas ligand interaction. Moreover, nitrated α-synuclein may activate CD4+ and CD4+ CD25+ T cells, and then they interact with microglia to initiate microgliosis and immune responses. Besides, some other events, such as mitochondrial dysfunction, protein aggregation, glutamate excitotoxicity, and loss of trophic factor support, may promote death of dopaminergic neurons. (For color version of this figure, the reader is referred to the online version of this chapter.)

6.1. Proinflammatory cytokines

Activated microglia and astrocytes produce several proinflammatory cytokines, including TNF-α, IL-1β, and IFN-γ (Przedborski, 2007). Their release creates an environment that mediates cell death (Herrera, Castano, Venero, Cano, & Machado, 2000).

6.1.1 Tumor necrosis factor

TNF, which is primarily secreted by microglia and astrocytes, is the main cytokine implicated in PD pathology (McCoy, Ruhn, Blesch, & Tansey, 2011; Smith et al., 2012). In PD brains, nigral midbrain dopaminergic neurons are extremely sensitive to TNF. Furthermore, elevated levels of TNF are seen in the CSF and postmortem brains of PD patients.

TNF signaling occurs via two distinct receptors: TNFR1, which is a canonical receptor for TNF signaling, and TNFR2, which activates the immune cascade, including anti-inflammatory and neuroprotective molecules to mediate a counter-regulatory response (Smith et al., 2012; Veroni et al., 2010). Upon binding of TNF to TNFR1, several signaling complexes are triggered that result in final activation of NFκB to drive the expression of antiapoptotic and proinflammatory proteins, such as Bcl-2, Bcl-xL, cytokines, and chemokines (Smith et al., 2012). TNF may also trigger apoptotic cell death through TNFR1, TNFR1-associated death domain adaptor protein (TRADD), and Fas-associated death domain protein (FADD), which subsequently activate initiator caspases, such as caspase-8 and -10, that ultimately induce cytochrome *c* release from mitochondria, causing neuronal apoptosis (Kraft, McPherson, & Harry, 2009; Smith et al., 2012). Hence, TNF and TNFRs are critical mediators of neurotoxic events because they are related to neuroinflammation, neuronal damage, and neuroprotection.

6.1.2 Interleukin-1

The IL-1 superfamily includes IL-1α, IL-1β, and endogenous IL-1 receptor antagonist (IL-1RA). IL-1α is a membrane-bound protein that takes part in paracrine and autocrine signaling, while IL-1β is a secreted protein (Smith et al., 2012). IL-1 family members are expressed by microglia and astrocytes during the neuroinflammatory process (Rappold & Tieu, 2010; Rivest, 2009). IL-1α and IL-1β signaling occurs mainly through type I IL-1R (IL-1RI). Upon binding of ligand to IL-1RI, several adaptors are recruited, namely, myeloid differentiation response gene 88 (MyD88) and TNF-associated factor 6 (TRAF6) (Smith et al., 2012). Then, the mitogen-activated protein kinase (MAPK) cascade is activated to further stimulate NFκB and activator protein-1 (AP-1) transcription factors that drive the transcription of IL-1-responsive genes. Moreover, IL-1RI activation may result in the hydrolysis of sphingomyelin to produce ceramide, which activates apoptotic pathways and causes changes in neuronal membrane physiology (Davis, Tabarean, Gaidarova, Behrens, & Bartfai, 2006; Hirsch et al., 2012).

6.1.3 Interferons

Interferons are glycoproteins and members of cytokine family that are secreted in response to pathogens or tumor cells. There are several types of interferons, such as IFN-α, IFN-β, and IFN-γ (Litteljohn et al., 2010).

Of those types of IFNs, IFN-γ was shown to exist in the neuroinflammatory environment of postmortem PD brains (Chakrabarty et al., 2011) (see Section 4.1 for details). IFN-γ is mainly produced by Th1 cells; however, activated microglia can also secrete IFN-γ (Kawanokuchi et al., 2006). According to epidemiological studies, IFN-γ levels are associated with PD risk and onset. Moreover, IFN-γ may activate oxidative and inflammatory enzyme systems along with TNF, namely, iNOS and NADPH oxidase (Litteljohn et al., 2010).

6.1.4 Fas/Fas ligand

In PD pathology, microglia mediate the recruitment of CD4+ T lymphocytes to dopaminergic neurons. Recruited and activated CD4+ T cells produce Fas ligand, as well as traditional proinflammatory molecules (Chen & Tansey, 2011). Fas ligand binds and activates Fas receptors on neuron, which induces apoptosis (Brochard et al., 2009). The phenomenon of Fas/Fas ligand activation is implicated in MPTP-induced PD animal models; Fas-deficient mice are resistant to MPTP because of the reduced effectiveness of recruited CD4+ T cells (Brochard et al., 2009). However, Fas ligand may also directly affect Fas receptor expressed on microglia, causing them to produce additional proinflammatory molecules. Hence, induction of the Fas/Fas ligand cascade is one of the mechanisms in PD pathology that directly or indirectly leads to dopaminergic neuron degeneration (Chen & Tansey, 2011; Hirsch & Hunot, 2009; Hirsch et al., 2012).

6.2. Cyclooxygenase-2

COX2, also known as prostaglandin-endoperoxide synthase 2, is an enzyme that is responsible for the production of prostaglandins during inflammation (Chen & Tansey, 2011). Pharmacological inactivation of COX2 is neuroprotective in MPTP-treated mice and 6-OHDA-treated rats (Teismann, Vila, et al., 2003). COX2 expression is inducible and occurs in neurons and glia in the CNS (Bartels & Leenders, 2010). In PD brains, COX expression is elevated in dopaminergic neurons and participates in dopaminergic neuron degeneration (Teismann, Tieu, et al., 2003). However, prostaglandin expression is not solely responsible for PD pathogenesis. During prostaglandin production, ROS are generated, which directly affect cell fate (Bartels & Leenders, 2010). Thus, COX is active in PD pathogenesis by indirectly affecting a cell's decision to undergo apoptosis.

6.3. Oxidative stress

In PD pathogenesis, microglial activation with gliosis results in an oxidative burst, which has toxic effects mediated by oxidative stress. During this burst, oxygen- and nitrogen-derived products, such as superoxide (O_2^-), nitric oxide (NO) species, and hypochlorous acid (HOCl), are released into the environment (Melo et al., 2011). For generation of oxidative stress, NADPH oxidase, iNOS, and MPO produce toxic amounts of the above-mentioned molecules (Chen & Tansey, 2011). In PD pathology, those biocatalytic systems are upregulated in the SN (Tansey & Goldberg, 2010; Tansey, McCoy, & Frank-Cannon, 2007). Their upregulation leads to increased production of oxidative stress-inducing molecules. Of these, O_2^- and NO free radicals react to generate highly toxic peroxynitrite $ONOO^-$ (Hirsch & Hunot, 2009; Przedborski, 2007; Tansey & Goldberg, 2010). Peroxynitrite may interact with several proteins inside the cell, such as α-*SYN* (Przedborski et al., 2001). Besides, in the presence of Fe^{2+} and Cu^{2+}, H_2O_2 may be converted to a hydroxy ($^{\bullet}OH$) radical via the Fenton reaction (Farooqui & Farooqui, 2011). Furthermore, nonreactive nitrite (NO_2^-) can be oxidized by MPO, resulting in the formation of a reactive form of nitrite, which can nitrosylate cellular proteins (van der Vliet, Eiserich, Halliwell, & Cross, 1997). The generation of an oxidative environment is a deleterious mechanism in PD pathology that leads to the death of vulnerable dopaminergic neurons (Chen & Tansey, 2011).

6.4. Glutamate excitotoxicity

Glutamate is an excitatory neurotransmitter amino acid that has direct, potent neurotoxic properties (Takeuchi, 2010). Under normal physiological conditions, excess glutamate is removed by astrocytes (Hanisch & Kettenmann, 2007). However, under pathological conditions, astroglial glutamate uptake is limited. Following the activation of microglia, released cytokines trigger microglia to secrete glutamate in an autocrine/paracrine manner. As the amount of extracellular glutamate increases in basal ganglia, excitotoxic cascades are initiated via either the *N*-methyl-D-aspartic acid (NMDA) receptor or the 2-amino-3-(5-methyl-3-oxo-1,2-oxazol-4-yl)propanoic acid (AMPA) receptor, which results in SNpc neurodegeneration (Caudle & Zhang, 2009). Moreover, SNpc may receive glutamatergic input from other regions of the basal ganglia. Additionally, NO may trigger glial cell-mediated glutamate release due to inhibition of mitochondrial respiration or via direct activation of vesicular exocytosis by modifying protein thiols (Brown, 2010).

Thus, microglia and astrocytes are responsible for controlling extracellular glutamate in the CNS environment to maintain homeostasis. Progressive excitotoxicity may result in immunoexcitotoxicity, which plays a central role in neurodegenerative diseases (Blaylock & Maroon, 2011).

6.5. Dysfunction of counter-regulatory and immunomodulatory mechanisms

The dysfunction of protective CNS mechanisms may result in the facilitation of neurodegeneration. In PD, dysfunction of glucocorticoid receptors (GRs) (Ros-Bernal et al., 2011) and nuclear receptor-related 1 (Nurr1) (Saijo et al., 2009) and downregulation of fractalkine (CX3CL1) (Noda et al., 2011) contribute to pathogenetic processes.

Glucocorticoids are released in response to stress and regulate inflammation to promote dopaminergic neuron survival (Ros-Bernal et al., 2011). In a recent study, GR levels analyzed in PD patients and MPTP-treated mice were found to be decreased in both (Ros-Bernal et al., 2011). When the authors specifically inactivated the GR gene in microglia and macrophages, MPTP-treated mice showed an increased loss of dopaminergic neurons and increased and persistent microglial activation.

Nurr1 is an orphan nuclear receptor that is responsible for the differentiation and maintenance of midbrain dopaminergic neurons (Bensinger & Tontonoz, 2009; Maguire-Zeiss & Federoff, 2010). Throughout life, Nurr1 regulates the expression of dopaminergic neuron-specific genes, namely, TH, DAT, vesicular monoamine transporter 2, and L-aromatic amino acid decarboxylase, which is the master regulator for dopaminergic neuron phenotype. Furthermore, Nurr1 controls the transcription of vasoactive intestinal protein (VIP) (Alavian, Scholz, & Simon, 2008; Federoff, 2009; Luo et al., 2007). A recent study suggests that Nurr1 is associated with inflammatory responses in microglia and astrocytes. That would imply that Nurr1 protects dopaminergic neurons from proinflammatory effects of microglia and astrocytes (Saijo et al., 2009). Hence, Nurr1, and therefore VIP, is responsible for protecting dopaminergic neurons from proinflammatory insults instigated by microglia and astrocytes.

CX3CL1, also known as fractalkine, is a chemokine that is secreted by neurons to repress inflammatory responses. Fractalkine mediates microglial phagocytosis of neuronal debris (Noda et al., 2011). In an environment without any proinflammatory molecules, fractalkine triggers the expression of antioxidant enzyme heme oxygenase-1 by activating Nrf2 (Noda et al., 2011). Moreover, fractalkine treatment of neuron–microglia cocultures diminishes glutamate-induced excitotoxic neuronal death. Thus, fractalkine is a very potent protector molecule that provides both phagocytic and neuroprotective effects.

CD200-CD200R signaling is responsible for restraining microglial activation (Hoarau et al., 2011). CD200R is expressed on microglia, but its ligand is present on neurons. In PD brains, CD200-CD200R signaling is impaired (Luo et al., 2010; Wang, Zhang, et al., 2011). Recent evidence suggests that CD200-CD200R signaling exacerbates microglial activation and neurodegeneration in the 6-OHDA-induced rat model of PD (Zhang, Phillips, et al., 2011).

6.6. Final death pathways

In the neurodegenerative process of PD, dopaminergic neurons die as a result of disturbances that are classified as proximal and distal events. In proximal events, α-*SYN* deposition and LB formation initiate apoptotic cell death signaling due to increased sensitivity to toxic insult (Levy, Malagelada, & Greene, 2009). On the other hand, distal events include the participation of a number of signal transduction pathways, such as c-Jun N-terminal kinase (JNK) signaling (Levy et al., 2009). In this signaling pathway, JNK phosphorylates several B-cell lymphoma 2 (Bcl-2) family member proteins in order to inhibit anti-inflammatory mechanisms and therefore activates proapoptotic Bcl-2-associated agonist of cell death (Bad) and Bcl2-like 11 (Bim) (Levy et al., 2009). Bim activation facilitates translocation of Bcl-2-associated X protein (Bax) to the mitochondria to release cytochrome *c* and induce apoptosis (Perier et al., 2007). Consequently, the death mechanism in PD occurs in a Bax-dependent manner with very distinct events. Therefore, Bcl2 family proteins may be utilized as useful biomarkers to assess cell death (Levy et al., 2009).

7. THERAPEUTIC IMPLICATIONS

7.1. Anti-inflammatory drugs

Neuroinflammation is an important mechanism implicated in PD pathogenesis. Microglial activation has been demonstrated in the SN of postmortem human brains and animal models of PD. Proinflammatory cytokine levels are increased in the SN and sera of PD patients and PD animal models. Many other inflammation-related molecules are also altered in PD. Evidence in

PD supports the hypothesis that anti-inflammatory therapy will be a useful new strategy in treating PD.

7.1.1 Nonsteroidal anti-inflammatory drugs

NSAIDs are a heterogeneous group of compounds that are commonly used as analgesics and antipyretics. NSAIDs mostly prevent COX activity. There are three main NSAID groups based on their COX inhibition capacities: nonselective COX inhibitors, selective COX1 inhibitors, and selective COX2 inhibitors (Esposito et al., 2007). Epidemiological and experimental evidence support the protective role of NSAIDs in parkinsonism. Cochrane reviews reported that ibuprofen (27%) reduced the risk of developing PD more than nonaspirin NSAIDs (13%) (Rees et al., 2011). NSAIDs influence other mechanisms that also mediate its neuroprotective effects, such as ROS production and the peroxisome proliferator-activated receptor γ (PPARγ) pathway (Asanuma, Nishibayashi-Asanuma, Miyazaki, Kohno, & Ogawa, 2001; Bernardo, Ajmone-Cat, Gasparini, Ongini, & Minghetti, 2005; Costa, Moutinho, Lima, & Fernandes, 2006). However, NSAIDs have not been tested in PD, so clinical studies are necessary to determine their exact role in PD treatment.

7.1.2 COX inhibitors

COX2 inhibitors show neuroprotective effects in MPTP-induced animal models of PD (Reksidler et al., 2007; Teismann & Ferger, 2001; Teismann, Tieu, et al., 2003). Nonselective COX inhibitors have serious side effects, particularly in the gastrointestinal system, which limits their long-term usage (Silverstein et al., 2000). Selective COX2 inhibitors do not have this side effect, but it has been reported that some COX2 inhibitors produce other serious adverse effects, including heart attack and stroke (White, 2007). New, safer anti-inflammatory drugs should be developed before the start of clinical studies with anti-inflammatory drugs.

7.1.3 Minocycline

The broad-spectrum antibiotic minocycline also shows anti-inflammatory effects (Kim & Suh, 2009). Minocycline provides neuroprotection in various animal models of PD, including 6-OHDA and MPTP (Du et al., 2001; Quintero et al., 2006). Rotenone toxicity is reduced by minocycline in parkin-null mice (Casarejos et al., 2006). In addition, G-protein inwardly rectifying potassium channel (*Girk2*) mutant mice shows inflammatory neurodegeneration in the SN, and minocycline attenuates the nigrostriatal dopaminergic neurodegeneration observed in this model (Peng et al., 2006). However, severe side effects manifested during clinical testing. A high proportion (23%) of patients prematurely discontinued minocycline in an 18-month phase II trial (NINSD NET-PD Investigators, 2008). This study does not support the safety of minocycline treatment of PD. Further studies are necessary to determine the safety of this drug and its treatment effect in PD.

7.1.4 PPARγ agonists

PPARγ is a type II nuclear receptor that regulates fatty acid storage and glucose metabolism (Whitehead, 2011). Many insulin-sensitizing drugs that are used in the treatment of diabetes target PPARγ (Libby & Plutzky, 2007). PPARγ also controls immune responses and could be a candidate drug to treat PD. Preclinical studies have shown that pioglitazone, a PPARγ agonist, prevents dopaminergic cell loss induced by MPTP in mice and monkeys (Dehmer, Heneka, Sastre, Dichgans, & Schulz, 2004; Quinn et al., 2008; Swanson et al., 2011). Several mechanisms, including MAO-B, NFκB and iNOS inhibition, and IκB alpha induction, mediate its neuroprotective effect in animal models of PD (Dehmer et al., 2004; Quinn et al., 2008). In addition, pioglitazone is protective in an inflammatory model of PD (Hunter, Choi, Ross, & Bing, 2008). PPARγ agonists were safely used in diabetic patients, so they may also be used in clinical trials for PD.

7.1.5 Adenosine antagonists

Adenosine is a nucleoside composed of adenine and ribose and acts as a signaling molecule. Four G-protein-coupled receptors (A1, A2A, A2B, and A3) mediate intracellular events stimulated by adenosine (Jenner et al., 2009). A2A receptor is mainly localized in the basal ganglia (Rodnitzky, 2012). Several preclinical studies support the neuroprotective effect of adenosine receptor antagonists in animal models of PD (Bibbiani et al., 2003; Hodgson et al., 2010; Kanda, Tashiro, Kuwana, & Jenner, 1998; Rose et al., 2006). Several clinical studies that employ adenosine receptor antagonists for PD treatment are ongoing. Phase II clinical trials suggested that istradefylline and preladenant reduce off-time and motor fluctuations in PD patients (Hauser et al., 2011; LeWitt et al., 2008; Pourcher et al., 2011). A phase III study with an adenosine receptor antagonist in PD treatment is currently underway.

7.1.6 Angiotensin II AT1 receptor inhibitors

The renin–angiotensin system (RAS) is the main modulator of salt and water homeostasis in the human body. Angiotensin II is the end product in RAS, and it is produced from angiotensin I by angiotensin-converting enzyme (Zucker & Ximmerman, 2011). Angiotensin II type 1 (AT1) and angiotensin II type 2 (AT2) receptors mediate its intracellular activities (Baltatu, Campos, & Bader, 2011). RAS is present in the brain and regulates many cellular functions, including inflammatory responses (Wright & Harding, 2011). Because angiotensin II is a proinflammatory molecule, angiotensin receptor antagonists such as losartan or angiotensin-converting enzyme inhibition show protective effects in animal models of PD (Grammatopoulos et al., 2007; Lopez-Real, Rey, Soto-Otero, Mendez-Alvarez, & Labandeira-Garcia, 2005; Munoz et al., 2006; Rey et al., 2007). ACE inhibitors have been used in patients with hypertension, and they could be tried in clinical studies of PD.

7.2. Parkinson's drugs with anti-inflammatory effects

7.2.1 Memantine

Memantine is an NMDA glutamate receptor blocker that is used to treat AD and PD. Recent studies suggest that it also has an anti-inflammatory action; memantine reduces serum and brain cytokine levels that are increased by chronic morphine treatment (Chen et al., 2011). In addition, memantine decreases inflammatory responses induced by LPS *in vitro* and *in vivo* (Rosi et al., 2006; Wu et al., 2009).

7.2.2 Amantadine

Amantadine is an antiviral agent that is also used in PD treatment. Various studies show that it has an anti-inflammatory function. Amantadine reduces T lymphocytes and alters the levels of several cytokines, including IL-2, TNF, and IFN-γ (Clark, Woodson, Winge, & Nagasawa, 1989; Kluter, Vieregge, Stolze, & Kirchner, 1995; Kubera et al., 2009; Tribl et al., 2001; Wandinger et al., 1999).

7.3. Immunotherapy

In neurodegenerative diseases like AD, PD, and ALS, neuroinflammation can be suppressed by immune-based therapeutic approaches that initiate innate and adaptive inflammation (Ha et al., 2012; Kosloski et al., 2010). From this perspective, drugs and anti-inflammatory effects of T cells can antagonize inflammatory actions that contribute to PD pathogenesis. Active and passive immunization can also be included in PD immunotherapy.

As stated in previous sections, T_{regs} can regulate T_{eff} activities by secreting immunosuppressive cytokines, namely, IL-10 and TGF-β, and by directly killing effector T cells (Huppa & Davis, 2003). In this manner, T_{reg}-induced attenuation of microglial inflammatory responses may have protective effects against nigrostriatal dopaminergic neurodegeneration (Kosloski et al., 2010). In a study using glatiramer acetate (GA), a synthetic 4-amino acid-length peptide used in MS patients, it was found that GA mediated the attenuation of microglial activation via induction of CD4+ T cells in an MPTP mouse model of PD (Laurie et al., 2007). Moreover, according to *in vitro* coculture studies with CD4+ T cells and microglia, nitrated α-SYN induced microglia apoptosis through Fas/FasL interactions (Reynolds et al., 2009a, 2009b). This evidence also supports the idea of adaptive immune dysregulation in PD. Furthermore, VIP can induce T_{reg}-mediated immunoregulatory responses (Kosloski et al., 2010). In a model using nitrated α-*SYN*, T_{regs} from VIP-treated mice suppressed T_{eff} responses, which protected dopaminergic neurons along the nigrostriatal axis (Reynolds et al., 2010). Furthermore, T_{regs} can be a source of neurotrophic factors. As a result of immunization with GA-specific T_{regs}, local GDNF expression was found to be elevated in an MPTP-induced PD model (Benner et al., 2004).

Vaccination is another possible treatment strategy for PD. In a study by Masliah et al., active immunization of mice with α-*SYN* resulted in the production of specific high-affinity antibodies (Masliah et al., 2005). In this model, extracellular α-*SYN* accumulation and the rate of neuronal death decreased. Antibodies produced after the introduction of α-*SYN* were shown to degrade α-SYN aggregates via lysosomal pathways (Schneeberger, Mandler, Mattner, & Schmidt, 2012). Moreover, AFFITOPE PD01 is a vaccine that was developed to treat patients with synucleinopathies (Schneeberger, Mandler, Mattner, & Schmidt, 2010). It targets PD-causing α-SYN aggregates by inducing production of antibodies against α-SYN (Schneeberger et al., 2012). On the other hand, a passive immunization strategy using antibodies can be a useful strategy for PD immunotherapy. Administration of monoclonal α-SYN antibody 9E4, which targets the C-terminal region of α-SYN, leads to the recovery of deficits induced by α-SYN accumulation (Masliah et al., 2011). Furthermore, the Bacille Calmette–Guérin (BCG) vaccine protects against striatal degeneration and prevents microglial activation in the SN of an MPTP mouse model of PD (Yong et al., 2011).

This evidence suggests that vaccination might be a viable option for treating PD. An increasing number of immunization studies are performed on animals but have not yet been translated into clinical practice. Therefore, more clinical trials should be initiated.

8. CONCLUSIONS

PD is the second most common age-related neurodegenerative disease associated with movement disorder. PD pathology is characterized by α-SYN deposition that causes degeneration of dopaminergic neurons in the SN, resulting in the disruption of dopaminergic pathways in the brain. Secondary to PD pathology, inflammation is a major mechanism that influences disease progression. Inflammatory reactions in PD are initially mediated by microglia and astrocytes with subsequent participation of the adaptive immune system.

In inflammatory responses in PD, activated microglia and astrocytes secrete proinflammatory cytokines that activate the transcription factor NFκB, which stimulates deleterious functions of microglia and astrocytes. On the other hand, activated microglia show phagocytose apoptotic cell debris and protein deposits. However, the precise inflammatory mechanisms involved in PD are not yet fully understood. To date, anti-inflammatory strategies to treat PD have only shown modest results.

Currently, palliative treatment approaches are used to treat PD. However, several novel treatment strategies are currently being developed, including vaccination. Although vaccination is still in infancy, preclinical studies suggest that it may be a potent way to treat complex diseases like PD. On the other hand, a combined treatment strategy using immunoregulatory, anti-inflammatory, and noninflammatory agents, such as antioxidants and neurotrophic factors, might also be useful PD treatments.

REFERENCES

Akiyama, H., & McGeer, P. L. (1989). Microglial response to 6-hydroxydopamine-induced substantia nigra lesions. *Brain Research*, *489*, 247–253.

Alam, M. M., & O'Neill, L. A. (2011). MicroRNAs and the resolution phase of inflammation in macrophages. *European Journal of Immunology*, *41*, 2482–2485.

Alavian, K. N., Scholz, C., & Simon, H. H. (2008). Transcriptional regulation of mesencephalic dopaminergic neurons: The full circle of life and death. *Movement Disorders: Official Journal of the Movement Disorder Society*, *23*, 319–328.

Allaman, I., Belanger, M., & Magistretti, P. J. (2011). Astrocyte-neuron metabolic relationships: For better and for worse. *Trends in Neurosciences*, *34*, 76–87.

Amor, S., Puentes, F., Baker, D., & van der Valk, P. (2010). Inflammation in neurodegenerative diseases. *Immunology, 129*, 154–169.

Antzoulatos, E., Jakowec, M. W., Petzinger, G. M., & Wood, R. I. (2010). Sex differences in motor behavior in the MPTP mouse model of Parkinson's disease. *Pharmacology, Biochemistry, and Behavior, 95*, 466–472.

Asanuma, M., Nishibayashi-Asanuma, S., Miyazaki, I., Kohno, M., & Ogawa, N. (2001). Neuroprotective effects of non-steroidal anti-inflammatory drugs by direct scavenging of nitric oxide radicals. *Journal of Neurochemistry, 76*, 1895–1904.

Baba, Y., Kuroiwa, A., Uitti, R. J., Wszolek, Z. K., & Yamada, T. (2005). Alterations of T-lymphocyte populations in Parkinson disease. *Parkinsonism & Related Disorders, 11*, 493–498.

Baltatu, O. C., Campos, L. A., & Bader, M. (2011). Local renin-angiotensin system and the brain—A continuous quest for knowledge. *Peptides, 32*, 1083–1086.

Bankiewicz, K. S., Oldfield, E. H., Chiueh, C. C., Doppman, J. L., Jacobowitz, D. M., & Kopin, I. J. (1986). Hemiparkinsonism in monkeys after unilateral internal carotid artery infusion of 1-methyl-4-phenyl-1,2,3,6-tetrahydropyridine (MPTP). *Life Sciences, 39*, 7–16.

Barcia, C., Ros, C. M., Annese, V., Gomez, A., Ros-Bernal, F., Aguado-Yera, D., et al. (2011). IFN-gamma signaling, with the synergistic contribution of TNF-alpha, mediates cell specific microglial and astroglial activation in experimental models of Parkinson's disease. *Cell Death & Disease, 2*, e142.

Bartels, A. L., & Leenders, K. L. (2009). Parkinson's disease: The syndrome, the pathogenesis and pathophysiology. *Cortex: A Journal Devoted to the Study of the Nervous System and Behavior, 45*, 915–921.

Bartels, A. L., & Leenders, K. L. (2010). Cyclooxygenase and neuroinflammation in Parkinson's disease neurodegeneration. *Current Neuropharmacology, 8*, 62–68.

Bartels, A. L., van Berckel, B. N., Lubberink, M., Luurtsema, G., Lammertsma, A. A., & Leenders, K. L. (2008). Blood-brain barrier P-glycoprotein function is not impaired in early Parkinson's disease. *Parkinsonism & Related Disorders, 14*, 505–508.

Bas, J., Calopa, M., Mestre, M., Mollevi, D. G., Cutillas, B., Ambrosio, S., et al. (2001). Lymphocyte populations in Parkinson's disease and in rat models of parkinsonism. *Journal of Neuroimmunology, 113*, 146–152.

Basso, M., Giraudo, S., Corpillo, D., Bergamasco, B., Lopiano, L., & Fasano, M. (2004). Proteome analysis of human substantia nigra in Parkinson's disease. *Proteomics, 4*, 3943–3952.

Benner, E. J., Mosley, R. L., Destache, C. J., Lewis, T. B., Jackson-Lewis, V., Gorantla, S., et al. (2004). Therapeutic immunization protects dopaminergic neurons in a mouse model of Parkinson's disease. *Proceedings of the National Academy of Sciences of the United States of America, 101*, 9435–9440.

Bensinger, S. J., & Tontonoz, P. (2009). A Nurr1 pathway for neuroprotection. *Cell, 137*, 26–28.

Beraud, D., & Maguire-Zeiss, K. A. (2012). Misfolded alpha-synuclein and Toll-like receptors: Therapeutic targets for Parkinson's disease. *Parkinsonism & Related Disorders, 18* (Suppl. 1), S17–S20.

Beraud, D., Twomey, M., Bloom, B., Mittereder, A., Ton, V., Neitzke, K., et al. (2011). Alpha-synuclein alters Toll-like receptor expression. *Frontiers in Neuroscience, 5*, 80.

Bernardo, A., Ajmone-Cat, M. A., Gasparini, L., Ongini, E., & Minghetti, L. (2005). Nuclear receptor peroxisome proliferator-activated receptor-gamma is activated in rat microglial cells by the anti-inflammatory drug HCT1026, a derivative of flurbiprofen. *Journal of Neurochemistry, 92*, 895–903.

Berry, C., La Vecchia, C., & Nicotera, P. (2010). Paraquat and Parkinson's disease. *Cell Death and Differentiation, 17*, 1115–1125.

Bialecka, M., Klodowska-Duda, G., Kurzawski, M., Slawek, J., Gorzkowska, A., Opala, G., et al. (2008). Interleukin-10 (IL10) and tumor necrosis factor alpha (TNF) gene polymorphisms in Parkinson's disease patients. *Parkinsonism & Related Disorders*, *14*, 636–640.
Bibbiani, F., Oh, J. D., Petzer, J. P., Castagnoli, N., Jr., Chen, J. F., Schwarzschild, M. A., et al. (2003). A2A antagonist prevents dopamine agonist-induced motor complications in animal models of Parkinson's disease. *Experimental Neurology*, *184*, 285–294.
Blandini, F. (2010). An update on the potential role of excitotoxicity in the pathogenesis of Parkinson's disease. *Functional Neurology*, *25*, 65–71.
Blaylock, R. L., & Maroon, J. (2011). Immunoexcitotoxicity as a central mechanism in chronic traumatic encephalopathy—A unifying hypothesis. *Surgical Neurology International*, *2*, 107.
Block, M. L., Zecca, L., & Hong, J. S. (2007). Microglia-mediated neurotoxicity: Uncovering the molecular mechanisms. *Nature Reviews. Neuroscience*, *8*, 57–69.
Blum-Degen, D., Muller, T., Kuhn, W., Gerlach, M., Przuntek, H., & Riederer, P. (1995). Interleukin-1 beta and interleukin-6 are elevated in the cerebrospinal fluid of Alzheimer's and de novo Parkinson's disease patients. *Neuroscience Letters*, *202*, 17–20.
Boka, G., Anglade, P., Wallach, D., Javoy-Agid, F., Agid, Y., & Hirsch, E. C. (1994). Immunocytochemical analysis of tumor necrosis factor and its receptors in Parkinson's disease. *Neuroscience Letters*, *172*, 151–154.
Bolin, L. M., Strycharska-Orczyk, I., Murray, R., Langston, J. W., & Di Monte, D. (2002). Increased vulnerability of dopaminergic neurons in MPTP-lesioned interleukin-6 deficient mice. *Journal of Neurochemistry*, *83*, 167–175.
Bornebroek, M., de Lau, L. M., Haag, M. D., Koudstaal, P. J., Hofman, A., Stricker, B. H., et al. (2007). Nonsteroidal anti-inflammatory drugs and the risk of Parkinson disease. *Neuroepidemiology*, *28*, 193–196.
Bove, J., & Perier, C. (2012). Neurotoxin-based models of Parkinson's disease. *Neuroscience*, *211*, 51–76.
Bower, J. H., Maraganore, D. M., Peterson, B. J., Ahlskog, J. E., & Rocca, W. A. (2006). Immunologic diseases, anti-inflammatory drugs, and Parkinson disease: A case-control study. *Neurology*, *67*, 494–496.
Brochard, V., Combadiere, B., Prigent, A., Laouar, Y., Perrin, A., Beray-Berthat, V., et al. (2009). Infiltration of CD4+ lymphocytes into the brain contributes to neurodegeneration in a mouse model of Parkinson disease. *The Journal of Clinical Investigation*, *119*, 182–192.
Brodacki, B., Staszewski, J., Toczylowska, B., Kozlowska, E., Drela, N., Chalimoniuk, M., et al. (2008). Serum interleukin (IL-2, IL-10, IL-6, IL-4), TNFalpha, and INFgamma concentrations are elevated in patients with atypical and idiopathic parkinsonism. *Neuroscience Letters*, *441*, 158–162.
Brown, G. C. (2010). Nitric oxide and neuronal death. *Nitric Oxide: Biology and Chemistry/Official Journal of the Nitric Oxide Society*, *23*, 153–165.
Calkins, M. J., Johnson, D. A., Townsend, J. A., Vargas, M. R., Dowell, J. A., Williamson, T. P., et al. (2009). The Nrf2/ARE pathway as a potential therapeutic target in neurodegenerative disease. *Antioxidants & Redox Signaling*, *11*, 497–508.
Calopa, M., Bas, J., Callen, A., & Mestre, M. (2010). Apoptosis of peripheral blood lymphocytes in Parkinson patients. *Neurobiology of Disease*, *38*, 1–7.
Cao, J. J., Li, K. S., & Shen, Y. Q. (2011). Activated immune cells in Parkinson's disease. *Journal of Neuroimmune Pharmacology: The Official Journal of the Society on NeuroImmune Pharmacology*, *6*, 323–329.
Casarejos, M. J., Menendez, J., Solano, R. M., Rodriguez-Navarro, J. A., Garcia de Yebenes, J., & Mena, M. A. (2006). Susceptibility to rotenone is increased in neurons from parkin null mice and is reduced by minocycline. *Journal of Neurochemistry*, *97*, 934–946.

Castano, A., Herrera, A. J., Cano, J., & Machado, A. (1998). Lipopolysaccharide intranigral injection induces inflammatory reaction and damage in nigrostriatal dopaminergic system. *Journal of Neurochemistry*, *70*, 1584–1592.
Caudle, W. M., & Zhang, J. (2009). Glutamate, excitotoxicity, and programmed cell death in Parkinson disease. *Experimental Neurology*, *220*, 230–233.
Chakrabarty, P., Ceballos-Diaz, C., Lin, W. L., Beccard, A., Jansen-West, K., McFarland, N. R., et al. (2011). Interferon-gamma induces progressive nigrostriatal degeneration and basal ganglia calcification. *Nature Neuroscience*, *14*, 694–696.
Chen, H., Jacobs, E., Schwarzschild, M. A., McCullough, M. L., Calle, E. E., Thun, M. J., et al. (2005). Nonsteroidal antiinflammatory drug use and the risk for Parkinson's disease. *Annals of Neurology*, *58*, 963–967.
Chen, X., & Tansey, M. G. (2011). The role of neuroinflammation in Parkinson's disease. In M. Alireza (Ed.), *Neuroinflammation* (pp. 403–421). London: Elsevier.
Chen, S. L., Tao, P. L., Chu, C. H., Chen, S. H., Wu, H. E., Tseng, L. F., Hong, J. S., & Lu, R. B. (2012). Low-dose memantine attenuated morphine addictive behavior through its anti-inflammation and neurotrophic effects in rats. *Journal of Neuroimmune Pharmacology: The Official Journal of the Society on NeuroImmune Pharmacology*, 7, 444–453.
Chen, H., Zhang, S. M., Hernan, M. A., Schwarzschild, M. A., Willett, W. C., Colditz, G. A., et al. (2003). Nonsteroidal anti-inflammatory drugs and the risk of Parkinson disease. *Archives of Neurology*, *60*, 1059–1064.
Chen, H., Zhang, S. M., Schwarzschild, M. A., Hernan, M. A., Willett, W. C., & Ascherio, A. (2004). Obesity and the risk of Parkinson's disease. *American Journal of Epidemiology*, *159*, 547–555.
Chiueh, C. C., Markey, S. P., Burns, R. S., Johannessen, J. N., Jacobowitz, D. M., & Kopin, I. J. (1984). Neurochemical and behavioral effects of 1-methyl-4-phenyl-1,2,3,6-tetrahydropyridine (MPTP) in rat, guinea pig, and monkey. *Psychopharmacology Bulletin*, *20*, 548–553.
Chung, Y. C., Ko, H. W., Bok, E., Park, E. S., Huh, S. H., Nam, J. H., et al. (2010). The role of neuroinflammation on the pathogenesis of Parkinson's disease. *BMB Reports*, *43*, 225–232.
Cicchetti, F., Lapointe, N., Roberge-Tremblay, A., Saint-Pierre, M., Jimenez, L., Ficke, B. W., et al. (2005). Systemic exposure to paraquat and maneb models early Parkinson's disease in young adult rats. *Neurobiology of Disease*, *20*, 360–371.
Clark, C., Woodson, M. M., Winge, V. B., & Nagasawa, H. T. (1989). The antiviral drug amantadine has a direct inhibitory effect on T-lymphocytes. *Immunopharmacology*, *18*, 195–204.
Collier, T. J., Kanaan, N. M., & Kordower, J. H. (2011). Ageing as a primary risk factor for Parkinson's disease: Evidence from studies of non-human primates. *Nature Reviews. Neuroscience*, *12*, 359–366.
Constantinescu, R., Andreasson, U., Li, S., Podust, V. N., Mattsson, N., Anckarsater, R., et al. (2010). Proteomic profiling of cerebrospinal fluid in parkinsonian disorders. *Parkinsonism & Related Disorders*, *16*, 545–549.
Coon, S., Stark, A., Peterson, E., Gloi, A., Kortsha, G., Pounds, J., et al. (2006). Whole-body lifetime occupational lead exposure and risk of Parkinson's disease. *Environmental Health Perspectives*, *114*, 1872–1876.
Cornejo Castro, E. M., Waak, J., Weber, S. S., Fiesel, F. C., Oberhettinger, P., Schutz, M., et al. (2010). Parkinson's disease-associated DJ-1 modulates innate immunity signaling in *Caenorhabditis elegans*. *Journal of Neural Transmission*, *117*, 599–604.
Corvol, J. C., Bonnet, C., Charbonnier-Beaupel, F., Bonnet, A. M., Fievet, M. H., Bellanger, A., et al. (2011). The COMT Val158Met polymorphism affects the response to entacapone in Parkinson's disease: A randomized crossover clinical trial. *Annals of Neurology*, *69*, 111–118.

Costa, D., Moutinho, L., Lima, J. L., & Fernandes, E. (2006). Antioxidant activity and inhibition of human neutrophil oxidative burst mediated by arylpropionic acid non-steroidal anti-inflammatory drugs. *Biological & Pharmaceutical Bulletin*, *29*, 1659–1670.
Cote, M., Drouin-Ouellet, J., Cicchetti, F., & Soulet, D. (2011). The critical role of the MyD88-dependent pathway in non-CNS MPTP-mediated toxicity. *Brain, Behavior, and Immunity*, *25*, 1143–1152.
Czeh, M., Gressens, P., & Kaindl, A. M. (2011). The yin and yang of microglia. *Developmental Neuroscience*, *33*, 199–209.
Damier, P., Hirsch, E. C., Zhang, P., Agid, Y., & Javoy-Agid, F. (1993). Glutathione peroxidase, glial cells and Parkinson's disease. *Neuroscience*, *52*, 1–6.
Davis, C. N., Tabarean, I., Gaidarova, S., Behrens, M. M., & Bartfai, T. (2006). IL-1beta induces a MyD88-dependent and ceramide-mediated activation of Src in anterior hypothalamic neurons. *Journal of Neurochemistry*, *98*, 1379–1389.
de Lau, L. M., & Breteler, M. M. (2006). Epidemiology of Parkinson's disease. *Lancet Neurology*, *5*, 525–535.
de Meira Santos Lima, M., Braga Reksidler, A., Marques Zanata, S., Bueno Machado, H., Tufik, S., & Vital, M. A. (2006). Different parkinsonism models produce a time-dependent induction of COX-2 in the substantia nigra of rats. *Brain Research*, *1101*, 117–125.
Deane, R., Bell, R. D., Sagare, A., & Zlokovic, B. V. (2009). Clearance of amyloid-beta peptide across the blood-brain barrier: Implication for therapies in Alzheimer's disease. *CNS & Neurological Disorders Drug Targets*, *8*, 16–30.
Dehmer, T., Heneka, M. T., Sastre, M., Dichgans, J., & Schulz, J. B. (2004). Protection by pioglitazone in the MPTP model of Parkinson's disease correlates with I kappa B alpha induction and block of NF kappa B and iNOS activation. *Journal of Neurochemistry*, *88*, 494–501.
Deleidi, M., Hallett, P. J., Koprich, J. B., Chung, C. Y., & Isacson, O. (2010). The Toll-like receptor-3 agonist polyinosinic: Polycytidylic acid triggers nigrostriatal dopaminergic degeneration. *The Journal of Neuroscience: The Official Journal of the Society for Neuroscience*, *30*, 16091–16101.
Desai Bradaric, B., Patel, A., Schneider, J. A., Carvey, P. M., & Hendey, B. (2012). Evidence for angiogenesis in Parkinson's disease, incidental Lewy body disease, and progressive supranuclear palsy. *Journal of Neural Transmission*, *119*, 59–71.
Desai, V. G., Feuers, R. J., Hart, R. W., & Ali, S. F. (1996). MPP(+)-induced neurotoxicity in mouse is age-dependent: Evidenced by the selective inhibition of complexes of electron transport. *Brain Research*, *715*, 1–8.
Dilger, R. N., & Johnson, R. W. (2008). Aging, microglial cell priming, and the discordant central inflammatory response to signals from the peripheral immune system. *Journal of Leukocyte Biology*, *84*, 932–939.
Dorsey, E. R., Constantinescu, R., Thompson, J. P., Biglan, K. M., Holloway, R. G., Kieburtz, K., et al. (2007). Projected number of people with Parkinson disease in the most populous nations, 2005 through 2030. *Neurology*, *68*, 384–386.
Double, K. L. (2012). Neuronal vulnerability in Parkinson's disease. *Parkinsonism & Related Disorders*, *18*(Suppl. 1), S52–S54.
Double, K. L., Reyes, S., Werry, E. L., & Halliday, G. M. (2010). Selective cell death in neurodegeneration: Why are some neurons spared in vulnerable regions? *Progress in Neurobiology*, *92*, 316–329.
Downer, E. J., Cowley, T. R., Lyons, A., Mills, K. H., Berezin, V., Bock, E., et al. (2010). A novel anti-inflammatory role of NCAM-derived mimetic peptide, FGL. *Neurobiology of Aging*, *31*, 118–128.
Drolet, R. E., Behrouz, B., Lookingland, K. J., & Goudreau, J. L. (2004). Mice lacking alpha-synuclein have an attenuated loss of striatal dopamine following prolonged chronic MPTP administration. *Neurotoxicology*, *25*, 761–769.

Du, Y., Ma, Z., Lin, S., Dodel, R. C., Gao, F., Bales, K. R., et al. (2001). Minocycline prevents nigrostriatal dopaminergic neurodegeneration in the MPTP model of Parkinson's disease. *Proceedings of the National Academy of Sciences of the United States of America, 98*, 14669–14674.

Dufek, M., Hamanova, M., Lokaj, J., Goldemund, D., Rektorova, I., Michalkova, Z., et al. (2009). Serum inflammatory biomarkers in Parkinson's disease. *Parkinsonism & Related Disorders, 15*, 318–320.

Elbaz, A., & Moisan, F. (2008). Update in the epidemiology of Parkinson's disease. *Current Opinion in Neurology, 21*, 454–460.

Elmer, L. W., & Bertoni, J. M. (2008). The increasing role of monoamine oxidase type B inhibitors in Parkinson's disease therapy. *Expert Opinion on Pharmacotherapy, 9*, 2759–2772.

Emmanouilidou, E., Elenis, D., Papasilekas, T., Stranjalis, G., Gerozissis, K., Ioannou, P. C., et al. (2011). Assessment of alpha-synuclein secretion in mouse and human brain parenchyma. *PLoS One, 6*, e22225.

Esposito, E., Di Matteo, V., Benigno, A., Pierucci, M., Crescimanno, G., & Di Giovanni, G. (2007). Non-steroidal anti-inflammatory drugs in Parkinson's disease. *Experimental Neurology, 205*, 295–312.

Etminan, M., Gill, S. S., & Samii, A. (2005). Intake of vitamin E, vitamin C, and carotenoids and the risk of Parkinson's disease: A meta-analysis. *Lancet Neurology, 4*, 362–365.

Farkas, E., De Jong, G. I., de Vos, R. A., Jansen Steur, E. N., & Luiten, P. G. (2000). Pathological features of cerebral cortical capillaries are doubled in Alzheimer's disease and Parkinson's disease. *Acta Neuropathologica, 100*, 395–402.

Farooqui, T., & Farooqui, A. A. (2011). Lipid-mediated oxidative stress and inflammation in the pathogenesis of Parkinson's disease. *Parkinson's Disease, 2011*, 247467.

Faucheux, B. A., Bonnet, A. M., Agid, Y., & Hirsch, E. C. (1999). Blood vessels change in the mesencephalon of patients with Parkinson's disease. *Lancet, 353*, 981–982.

Federoff, H. J. (2009). Nur(R1)turing a notion on the etiopathogenesis of Parkinson's disease. *Neurotoxicity Research, 16*, 261–270.

Finehout, E. J., Franck, Z., & Lee, K. H. (2005). Complement protein isoforms in CSF as possible biomarkers for neurodegenerative disease. *Disease Markers, 21*, 93–101.

Fiszer, U., Mix, E., Fredrikson, S., Kostulas, V., Olsson, T., & Link, H. (1994). Gamma delta+ T cells are increased in patients with Parkinson's disease. *Journal of the Neurological Sciences, 121*, 39–45.

Foley, P. B. (2009). Encephalitis lethargica and influenza. I. The role of the influenza virus in the influenza pandemic of 1918/1919. *Journal of Neural Transmission, 116*, 143–150.

Frank, M. G., Barrientos, R. M., Watkins, L. R., & Maier, S. F. (2010). Aging sensitizes rapidly isolated hippocampal microglia to LPS ex vivo. *Journal of Neuroimmunology, 226*, 181–184.

Freeman, M. R. (2010). Specification and morphogenesis of astrocytes. *Science, 330*, 774–778.

Fukushima, W., Miyake, Y., Tanaka, K., Sasaki, S., Kiyohara, C., Tsuboi, Y., et al. (2010). Alcohol drinking and risk of Parkinson's disease: A case-control study in Japan. *BMC Neurology, 10*, 111.

Fulceri, F., Biagioni, F., Lenzi, P., Falleni, A., Gesi, M., Ruggieri, S., et al. (2006). Nigrostriatal damage with 6-OHDA: Validation of routinely applied procedures. *Annals of the New York Academy of Sciences, 1074*, 344–348.

Furuya, T., Hayakawa, H., Yamada, M., Yoshimi, K., Hisahara, S., Miura, M., et al. (2004). Caspase-11 mediates inflammatory dopaminergic cell death in the 1-methyl-4-phenyl-1,2,3,6-tetrahydropyridine mouse model of Parkinson's disease. *The Journal of Neuroscience: The Official Journal of the Society for Neuroscience, 24*, 1865–1872.

Fuxe, K. G., Tarakanov, A. O., Goncharova, L. B., & Agnati, L. F. (2008). A new road to neuroinflammation in Parkinson's disease? *Brain Research Reviews, 58*, 453–458.

Gagne, J. J., & Power, M. C. (2010). Anti-inflammatory drugs and risk of Parkinson disease: A meta-analysis. *Neurology*, *74*, 995–1002.

Galea, I., Bechmann, I., & Perry, V. H. (2007). What is immune privilege (not)? *Trends in Immunology*, *28*, 12–18.

Gao, X., Chen, H., Schwarzschild, M. A., & Ascherio, A. (2011a). Use of ibuprofen and risk of Parkinson disease. *Neurology*, *76*, 863–869.

Gao, H. M., & Hong, J. S. (2008). Why neurodegenerative diseases are progressive: Uncontrolled inflammation drives disease progression. *Trends in Immunology*, *29*, 357–365.

Gao, H. M., Kotzbauer, P. T., Uryu, K., Leight, S., Trojanowski, J. Q., & Lee, V. M. (2008). Neuroinflammation and oxidation/nitration of alpha-synuclein linked to dopaminergic neurodegeneration. *The Journal of Neuroscience: The Official Journal of the Society for Neuroscience*, *28*, 7687–7698.

Gao, H. M., Zhang, F., Zhou, H., Kam, W., Wilson, B., & Hong, J. S. (2011b). Neuroinflammation and alpha-synuclein dysfunction potentiate each other, driving chronic progression of neurodegeneration in a mouse model of Parkinson's disease. *Environmental Health Perspectives*, *119*, 807–814.

Gardet, A., Benita, Y., Li, C., Sands, B. E., Ballester, I., Stevens, C., et al. (2010). LRRK2 is involved in the IFN-gamma response and host response to pathogens. *Journal of Immunology*, *185*, 5577–5585.

Gendelman, H. E., & Appel, S. H. (2011). Neuroprotective activities of regulatory T cells. *Trends in Molecular Medicine*, *17*, 687–688.

Geng, Y., Fang, M., Wang, J., Yu, H., Hu, Z., Yew, D. T., et al. (2011). Triptolide down-regulates COX-2 expression and PGE2 release by suppressing the activity of NF-kappaB and MAP kinases in lipopolysaccharide-treated PC12 cells. *Phytotherapy Research*, *26*, 337–343.

Gerhard, A., Pavese, N., Hotton, G., Turkheimer, F., Es, M., Hammers, A., et al. (2006). In vivo imaging of microglial activation with [11C](R)-PK11195 PET in idiopathic Parkinson's disease. *Neurobiology of Disease*, *21*, 404–412.

Giasson, B. I., Duda, J. E., Quinn, S. M., Zhang, B., Trojanowski, J. Q., & Lee, V. M. (2002). Neuronal alpha-synucleinopathy with severe movement disorder in mice expressing A53T human alpha-synuclein. *Neuron*, *34*, 521–533.

Glass, C. K., Saijo, K., Winner, B., Marchetto, M. C., & Gage, F. H. (2010). Mechanisms underlying inflammation in neurodegeneration. *Cell*, *140*, 918–934.

Gomide, V., Bibancos, T., & Chadi, G. (2005). Dopamine cell morphology and glial cell hypertrophy and process branching in the nigrostriatal system after striatal 6-OHDA analyzed by specific sterological tools. *The International Journal of Neuroscience*, *115*, 557–582.

Gorell, J. M., Johnson, C. C., Rybicki, B. A., Peterson, E. L., Kortsha, G. X., Brown, G. G., et al. (1999). Occupational exposure to manganese, copper, lead, iron, mercury and zinc and the risk of Parkinson's disease. *Neurotoxicology*, *20*, 239–247.

Graeber, M. B., Li, W., & Rodriguez, M. L. (2011). Role of microglia in CNS inflammation. *FEBS Letters*, *585*, 3798–3805.

Graeber, M. B., & Streit, W. J. (2010). Microglia: Biology and pathology. *Acta Neuropathologica*, *119*, 89–105.

Grammatopoulos, T. N., Jones, S. M., Ahmadi, F. A., Hoover, B. R., Snell, L. D., Skoch, J., et al. (2007). Angiotensin type 1 receptor antagonist losartan, reduces MPTP-induced degeneration of dopaminergic neurons in substantia nigra. *Molecular Neurodegeneration*, *2*, 1.

Greenamyre, J. T., Cannon, J. R., Drolet, R., & Mastroberardino, P. G. (2010). Lessons from the rotenone model of Parkinson's disease. *Trends in Pharmacological Sciences*, *31*, 141–142 author reply 142–143.

Greene, J. C., Whitworth, A. J., Andrews, L. A., Parker, T. J., & Pallanck, L. J. (2005). Genetic and genomic studies of Drosophila parkin mutants implicate oxidative stress and innate immune responses in pathogenesis. *Human Molecular Genetics*, *14*, 799–811.

Griffiths, M. R., Gasque, P., & Neal, J. W. (2010). The regulation of the CNS innate immune response is vital for the restoration of tissue homeostasis (repair) after acute brain injury: A brief review. *International Journal of Inflammation, 2010*, 151097.
Grunblatt, E., Mandel, S., Maor, G., & Youdim, M. B. (2001). Gene expression analysis in N-methyl-4-phenyl-1,2,3,6-tetrahydropyridine mice model of Parkinson's disease using cDNA microarray: Effect of R-apomorphine. *Journal of Neurochemistry, 78*, 1–12.
Grundemann, J., Schlaudraff, F., Haeckel, O., & Liss, B. (2008). Elevated alpha-synuclein mRNA levels in individual UV-laser-microdissected dopaminergic substantia nigra neurons in idiopathic Parkinson's disease. *Nucleic Acids Research, 36*, e38.
Gupta, M., Gupta, B. K., Thomas, R., Bruemmer, V., Sladek, J. R., Jr., & Felten, D. L. (1986). Aged mice are more sensitive to 1-methyl-4-phenyl-1,2,3,6-tetrahydropyridine treatment than young adults. *Neuroscience Letters, 70*, 326–331.
Ha, D., Stone, D. K., Mosley, R. L., & Gendelman, H. E. (2012). Immunization strategies for Parkinson's disease. *Parkinsonism & Related Disorders, 18*(Suppl. 1), S218–S221.
Hakansson, A., Westberg, L., Nilsson, S., Buervenich, S., Carmine, A., Holmberg, B., et al. (2005). Interaction of polymorphisms in the genes encoding interleukin-6 and estrogen receptor beta on the susceptibility to Parkinson's disease. *American Journal of Medical Genetics. Part B: Neuropsychiatric Genetics: The Official Publication of the International Society of Psychiatric Genetics, 133B*, 88–92.
Hakimi, M., Selvanantham, T., Swinton, E., Padmore, R. F., Tong, Y., Kabbach, G., et al. (2011). Parkinson's disease-linked LRRK2 is expressed in circulating and tissue immune cells and upregulated following recognition of microbial structures. *Journal of Neural Transmission, 118*, 795–808.
Halliday, G. M., & Stevens, C. H. (2011). Glia: Initiators and progressors of pathology in Parkinson's disease. *Movement Disorders: Official Journal of the Movement Disorder Society, 26*, 6–17.
Hamby, M. E., & Sofroniew, M. V. (2010). Reactive astrocytes as therapeutic targets for CNS disorders. *Neurotherapeutics: The Journal of the American Society for Experimental NeuroTherapeutics, 7*, 494–506.
Hamill, C. E., Caudle, W. M., Richardson, J. R., Yuan, H., Pennell, K. D., Greene, J. G., et al. (2007). Exacerbation of dopaminergic terminal damage in a mouse model of Parkinson's disease by the G-protein-coupled receptor protease-activated receptor 1. *Molecular Pharmacology, 72*, 653–664.
Hamza, T. H., Zabetian, C. P., Tenesa, A., Laederach, A., Montimurro, J., Yearout, D., et al. (2010). Common genetic variation in the HLA region is associated with late-onset sporadic Parkinson's disease. *Nature Genetics, 42*, 781–785.
Hancock, D. B., Martin, E. R., Stajich, J. M., Jewett, R., Stacy, M. A., Scott, B. L., et al. (2007). Smoking, caffeine, and nonsteroidal anti-inflammatory drugs in families with Parkinson disease. *Archives of Neurology, 64*, 576–580.
Hanisch, U. K., & Kettenmann, H. (2007). Microglia: Active sensor and versatile effector cells in the normal and pathologic brain. *Nature Neuroscience, 10*, 1387–1394.
Hanke, M. L., & Kielian, T. (2011). Toll-like receptors in health and disease in the brain: Mechanisms and therapeutic potential. *Clinical Science (London, England), 121*, 367–387.
Harms, A. S., Barnum, C. J., Ruhn, K. A., Varghese, S., Trevino, I., Blesch, A., et al. (2011). Delayed dominant-negative TNF gene therapy halts progressive loss of nigral dopaminergic neurons in a rat model of Parkinson's disease. *Molecular Therapy: The Journal of the American Society of Gene Therapy, 19*, 46–52.
Hauser, R. A., Cantillon, M., Pourcher, E., Micheli, F., Mok, V., Onofrj, M., et al. (2011). Preladenant in patients with Parkinson's disease and motor fluctuations: A phase 2, double-blind, randomised trial. *Lancet Neurology, 10*, 221–229.

Helmy, A., De Simoni, M. G., Guilfoyle, M. R., Carpenter, K. L., & Hutchinson, P. J. (2011). Cytokines and innate inflammation in the pathogenesis of human traumatic brain injury. *Progress in Neurobiology, 95*, 352–372.
Hernan, M. A., Logroscino, G., & Garcia Rodriguez, L. A. (2006). Nonsteroidal anti-inflammatory drugs and the incidence of Parkinson disease. *Neurology, 66*, 1097–1099.
Hernandez-Romero, M. C., Arguelles, S., Villaran, R. F., de Pablos, R. M., Delgado-Cortes, M. J., Santiago, M., et al. (2008). Simvastatin prevents the inflammatory process and the dopaminergic degeneration induced by the intranigral injection of lipopolysaccharide. *Journal of Neurochemistry, 105*, 445–459.
Herrera, A. J., Castano, A., Venero, J. L., Cano, J., & Machado, A. (2000). The single intranigral injection of LPS as a new model for studying the selective effects of inflammatory reactions on dopaminergic system. *Neurobiology of Disease*, 7, 429–447.
Hirsch, E. C., & Hunot, S. (2009). Neuroinflammation in Parkinson's disease: A target for neuroprotection? *Lancet Neurology, 8*, 382–397.
Hirsch, E. C., Vyas, S., & Hunot, S. (2012). Neuroinflammation in Parkinson's disease. *Parkinsonism & Related Disorders, 18*(Suppl. 1), S210–S212.
Hoarau, J. J., Krejbich-Trotot, P., Jaffar-Bandjee, M. C., Das, T., Thon-Hon, G. V., Kumar, S., et al. (2011). Activation and control of CNS innate immune responses in health and diseases: A balancing act finely tuned by neuroimmune regulators (NIReg). *CNS & Neurological Disorders Drug Targets, 10*, 25–43.
Hodgson, R. A., Bedard, P. J., Varty, G. B., Kazdoba, T. M., Di Paolo, T., Grzelak, M. E., et al. (2010). Preladenant, a selective A(2A) receptor antagonist, is active in primate models of movement disorders. *Experimental Neurology, 225*, 384–390.
Hofmann, K. W., Schuh, A. F., Saute, J., Townsend, R., Fricke, D., Leke, R., et al. (2009). Interleukin-6 serum levels in patients with Parkinson's disease. *Neurochemical Research, 34*, 1401–1404.
Horstman, L. L., Jy, W., Ahn, Y. S., Maghzi, A. H., Etemadifar, M., Alexander, J. S., et al. (2012). Complement in neurobiology. *Frontiers in Bioscience: A Journal and Virtual Library, 17*, 2921–2960.
Hu, G., Antikainen, R., Jousilahti, P., Kivipelto, M., & Tuomilehto, J. (2008). Total cholesterol and the risk of Parkinson disease. *Neurology, 70*, 1972–1979.
Hunter, R. L., Choi, D. Y., Ross, S. A., & Bing, G. (2008). Protective properties afforded by pioglitazone against intrastriatal LPS in Sprague-Dawley rats. *Neuroscience Letters, 432*, 198–201.
Huppa, J. B., & Davis, M. M. (2003). T-cell-antigen recognition and the immunological synapse. *Nature Reviews. Immunology, 3*, 973–983.
Imamura, K., Hishikawa, N., Sawada, M., Nagatsu, T., Yoshida, M., & Hashizume, Y. (2003). Distribution of major histocompatibility complex class II-positive microglia and cytokine profile of Parkinson's disease brains. *Acta Neuropathologica, 106*, 518–526.
Iravani, M. M., Leung, C. C., Sadeghian, M., Haddon, C. O., Rose, S., & Jenner, P. (2005). The acute and the long-term effects of nigral lipopolysaccharide administration on dopaminergic dysfunction and glial cell activation. *The European Journal of Neuroscience, 22*, 317–330.
Jackson-Lewis, V., Blesa, J., & Przedborski, S. (2012). Animal models of Parkinson's disease. *Parkinsonism & Related Disorders, 18*(Suppl. 1), S183–S185.
Jang, H., Boltz, D. A., Webster, R. G., & Smeyne, R. J. (2009). Viral parkinsonism. *Biochimica et Biophysica Acta, 1792*, 714–721.
Jellinger, K. A. (2009). Recent advances in our understanding of neurodegeneration. *Journal of Neural Transmission, 116*, 1111–1162.
Jenner, P., Mori, A., Hauser, R., Morelli, M., Fredholm, B. B., & Chen, J. F. (2009). Adenosine, adenosine A 2A antagonists, and Parkinson's disease. *Parkinsonism & Related Disorders, 15*, 406–413.

Jin, J., Shie, F. S., Liu, J., Wang, Y., Davis, J., Schantz, A. M., et al. (2007). Prostaglandin E2 receptor subtype 2 (EP2) regulates microglial activation and associated neurotoxicity induced by aggregated alpha-synuclein. *Journal of Neuroinflammation*, *4*, 2.

Johannessen, J. N. (1991). A model of chronic neurotoxicity: Long-term retention of the neurotoxin 1-methyl-4-phenylpyridinium (MPP+) within catecholaminergic neurons. *Neurotoxicology*, *12*, 285–302.

Jurgens, H. A., & Johnson, R. W. (2012). Dysregulated neuronal-microglial cross-talk during aging, stress and inflammation. *Experimental Neurology*, *233*, 40–48.

Kacimi, R., Giffard, R. G., & Yenari, M. A. (2011). Endotoxin-activated microglia injure brain derived endothelial cells via NF-kappaB, JAK-STAT and JNK stress kinase pathways. *Journal of Inflammation (London)*, *8*, 7.

Kanda, T., Tashiro, T., Kuwana, Y., & Jenner, P. (1998). Adenosine A2A receptors modify motor function in MPTP-treated common marmosets. *NeuroReport*, *9*, 2857–2860.

Kawanokuchi, J., Mizuno, T., Takeuchi, H., Kato, H., Wang, J., Mitsuma, N., et al. (2006). Production of interferon-gamma by microglia. *Multiple Sclerosis*, *12*, 558–564.

Khandelwal, P. J., Herman, A. M., & Moussa, C. E. (2011). Inflammation in the early stages of neurodegenerative pathology. *Journal of Neuroimmunology*, *238*, 1–11.

Kim, Y. S., Choi, D. H., Block, M. L., Lorenzl, S., Yang, L., Kim, Y. J., et al. (2007). A pivotal role of matrix metalloproteinase-3 activity in dopaminergic neuronal degeneration via microglial activation. *The FASEB Journal: Official Publication of the Federation of American Societies for Experimental Biology*, *21*, 179–187.

Kim, E. M., & Hwang, O. (2011). Role of matrix metalloproteinase-3 in neurodegeneration. *Journal of Neurochemistry*, *116*, 22–32.

Kim, J. H., Lee, H. W., Hwang, J., Kim, J., Lee, M. J., Han, H. S., et al. (2011). Microglia-inhibiting activity of Parkinson's disease drug amantadine. *Neurobiology of Aging*, (in press).

Kim, W. G., Mohney, R. P., Wilson, B., Jeohn, G. H., Liu, B., & Hong, J. S. (2000). Regional difference in susceptibility to lipopolysaccharide-induced neurotoxicity in the rat brain: Role of microglia. *The Journal of Neuroscience: The Official Journal of the Society for Neuroscience*, *20*, 6309–6316.

Kim, H. S., & Suh, Y. H. (2009). Minocycline and neurodegenerative diseases. *Behavioural Brain Research*, *196*, 168–179.

Kiyohara, C., Miyake, Y., Koyanagi, M., Fujimoto, T., Shirasawa, S., Tanaka, K., et al. (2011). APOE and CYP2E1 polymorphisms, alcohol consumption, and Parkinson's disease in a Japanese population. *Journal of Neural Transmission*, *118*, 1335–1344.

Kluter, H., Vieregge, P., Stolze, H., & Kirchner, H. (1995). Defective production of interleukin-2 in patients with idiopathic Parkinson's disease. *Journal of the Neurological Sciences*, *133*, 134–139.

Knott, C., Stern, G., & Wilkin, G. P. (2000). Inflammatory regulators in Parkinson's disease: iNOS, lipocortin-1, and cyclooxygenases-1 and -2. *Molecular and Cellular Neurosciences*, *16*, 724–739.

Kofler, J., & Wiley, C. A. (2011). Microglia: Key innate immune cells of the brain. *Toxicologic Pathology*, *39*, 103–114.

Kofuji, P., & Newman, E. A. (2004). Potassium buffering in the central nervous system. *Neuroscience*, *129*, 1045–1056.

Kosloski, L. M., Ha, D. M., Hutter, J. A., Stone, D. K., Pichler, M. R., Reynolds, A. D., et al. (2010). Adaptive immune regulation of glial homeostasis as an immunization strategy for neurodegenerative diseases. *Journal of Neurochemistry*, *114*, 1261–1276.

Kraft, A. D., McPherson, C. A., & Harry, G. J. (2009). Heterogeneity of microglia and TNF signaling as determinants for neuronal death or survival. *Neurotoxicology*, *30*, 785–793.

Kruger, R., Hardt, C., Tschentscher, F., Jackel, S., Kuhn, W., Muller, T., et al. (2000). Genetic analysis of immunomodulating factors in sporadic Parkinson's disease. *Journal of Neural Transmission*, *107*, 553–562.

Kubera, M., Maes, M., Budziszewska, B., Basta-Kaim, A., Leskiewicz, M., Grygier, B., et al. (2009). Inhibitory effects of amantadine on the production of pro-inflammatory cytokines by stimulated in vitro human blood. *Pharmacological Reports*, *61*, 1105–1112.
Lampe, J. B., Gossrau, G., Herting, B., Kempe, A., Sommer, U., Fussel, M., et al. (2003). HLA typing and Parkinson's disease. *European Neurology*, *50*, 64–68.
Langston, J. W., Forno, L. S., Rebert, C. S., & Irwin, I. (1984). Selective nigral toxicity after systemic administration of 1-methyl-4-phenyl-1,2,5,6-tetrahydropyrine (MPTP) in the squirrel monkey. *Brain Research*, *292*, 390–394.
Lau, Y. S., Patki, G., Das-Panja, K., Le, W. D., & Ahmad, S. O. (2011). Neuroprotective effects and mechanisms of exercise in a chronic mouse model of Parkinson's disease with moderate neurodegeneration. *The European Journal of Neuroscience*, *33*, 1264–1274.
Laurie, C., Reynolds, A., Coskun, O., Bowman, E., Gendelman, H. E., & Mosley, R. L. (2007). CD4+ T cells from Copolymer-1 immunized mice protect dopaminergic neurons in the 1-methyl-4-phenyl-1,2,3,6-tetrahydropyridine model of Parkinson's disease. *Journal of Neuroimmunology*, *183*, 60–68.
Lawson, L. J., Perry, V. H., Dri, P., & Gordon, S. (1990). Heterogeneity in the distribution and morphology of microglia in the normal adult mouse brain. *Neuroscience*, *39*, 151–170.
Lee, S. J. (2008). Origins and effects of extracellular alpha-synuclein: Implications in Parkinson's disease. *Journal of Molecular Neuroscience*, *34*, 17–22.
Lee, H. J., Kim, C., & Lee, S. J. (2010a). Alpha-synuclein stimulation of astrocytes: Potential role for neuroinflammation and neuroprotection. *Oxidative Medicine and Cellular Longevity*, *3*, 283–287.
Lee, M., Schwab, C., & McGeer, P. L. (2011). Astrocytes are GABAergic cells that modulate microglial activity. *Glia*, *59*, 152–165.
Lee, E. J., Woo, M. S., Moon, P. G., Baek, M. C., Choi, I. Y., Kim, W. K., et al. (2010b). Alpha-synuclein activates microglia by inducing the expressions of matrix metalloproteinases and the subsequent activation of protease-activated receptor-1. *Journal of Immunology*, *185*, 615–623.
Lees, A. J., Hardy, J., & Revesz, T. (2009). Parkinson's disease. *Lancet*, *373*, 2055–2066.
Levesque, S., Wilson, B., Gregoria, V., Thorpe, L. B., Dallas, S., Polikov, V. S., et al. (2010). Reactive microgliosis: Extracellular micro-calpain and microglia-mediated dopaminergic neurotoxicity. *Brain: A Journal of Neurology*, *133*, 808–821.
Levy, O. A., Malagelada, C., & Greene, L. A. (2009). Cell death pathways in Parkinson's disease: Proximal triggers, distal effectors, and final steps. *Apoptosis: An International Journal on Programmed Cell Death*, *14*, 478–500.
LeWitt, P. A., Guttman, M., Tetrud, J. W., Tuite, P. J., Mori, A., Chaikin, P., et al. (2008). Adenosine A2A receptor antagonist istradefylline (KW-6002) reduces "off" time in Parkinson's disease: A double-blind, randomized, multicenter clinical trial (6002-US-005). *Annals of Neurology*, *63*, 295–302.
Li, Z., Melandri, F., Berdo, I., Jansen, M., Hunter, L., Wright, S., et al. (2004). Delta12-Prostaglandin J2 inhibits the ubiquitin hydrolase UCH-L1 and elicits ubiquitin-protein aggregation without proteasome inhibition. *Biochemical and Biophysical Research Communications*, *319*, 1171–1180.
Libby, P., & Plutzky, J. (2007). Inflammation in diabetes mellitus: Role of peroxisome proliferator-activated receptor-alpha and peroxisome proliferator-activated receptor-gamma agonists. *The American Journal of Cardiology*, *99*, 27B–40B.
Lim, S. Y., & Lang, A. E. (2010). The nonmotor symptoms of Parkinson's disease—An overview. *Movement Disorders: Official Journal of the Movement Disorder Society*, *25*(Suppl. 1), S123–S130.
Lin, J. J., Chen, C. H., Yueh, K. C., Chang, C. Y., & Lin, S. Z. (2006). A CD14 monocyte receptor polymorphism and genetic susceptibility to Parkinson's disease for females. *Parkinsonism & Related Disorders*, *12*, 9–13.

Litteljohn, D., Mangano, E., Clarke, M., Bobyn, J., Moloney, K., & Hayley, S. (2010). Inflammatory mechanisms of neurodegeneration in toxin-based models of Parkinson's disease. *Parkinson's Disease, 2011*, 713517.

Liu, M., & Bing, G. (2011). Lipopolysaccharide animal models for Parkinson's disease. *Parkinson's Disease, 2011*, 327089.

Liu, X., Cheng, R., Verbitsky, M., Kisselev, S., Browne, A., Mejia-Sanatana, H., et al. (2011a). Genome-wide association study identifies candidate genes for Parkinson's disease in an Ashkenazi Jewish population. *BMC Medical Genetics, 12*, 104.

Liu, W., Tang, Y., & Feng, J. (2011b). Cross talk between activation of microglia and astrocytes in pathological conditions in the central nervous system. *Life Sciences, 89*, 141–146.

Liu, J. P., Tang, Y., Zhou, S., Toh, B. H., McLean, C., & Li, H. (2010a). Cholesterol involvement in the pathogenesis of neurodegenerative diseases. *Molecular and Cellular Neurosciences, 43*, 33–42.

Liu, J., Wang, M. W., Gu, P., Ma, Q. Y., Wang, Y. Y., Geng, Y., et al. (2010b). Microglial activation and age-related dopaminergic neurodegeneration in MPTP-treated SAMP8 mice. *Brain Research, 1345*, 213–220.

Lopez-Real, A., Rey, P., Soto-Otero, R., Mendez-Alvarez, E., & Labandeira-Garcia, J. L. (2005). Angiotensin-converting enzyme inhibition reduces oxidative stress and protects dopaminergic neurons in a 6-hydroxydopamine rat model of Parkinsonism. *Journal of Neuroscience Research, 81*, 865–873.

Lull, M. E., & Block, M. L. (2010). Microglial activation and chronic neurodegeneration. *Neurotherapeutics: The Journal of the American Society for Experimental NeuroTherapeutics*, 7, 354–365.

Luo, Y., Henricksen, L. A., Giuliano, R. E., Prifti, L., Callahan, L. M., & Federoff, H. J. (2007). VIP is a transcriptional target of Nurr1 in dopaminergic cells. *Experimental Neurology, 203*, 221–232.

Luo, X. G., Zhang, J. J., Zhang, C. D., Liu, R., Zheng, L., Wang, X. J., et al. (2010). Altered regulation of CD200 receptor in monocyte-derived macrophages from individuals with Parkinson's disease. *Neurochemical Research, 35*, 540–547.

Lynch, M. A., & Mills, K. H. (2012). Immunology meets neuroscience—Opportunities for immune intervention in neurodegenerative diseases. *Brain, Behavior, and Immunity, 26*, 1–10.

Maetzler, W., Berg, D., Schalamberidze, N., Melms, A., Schott, K., Mueller, J. C., et al. (2007). Osteopontin is elevated in Parkinson's disease and its absence leads to reduced neurodegeneration in the MPTP model. *Neurobiology of Disease, 25*, 473–482.

Maguire-Zeiss, K. A., & Federoff, H. J. (2010). Future directions for immune modulation in neurodegenerative disorders: Focus on Parkinson's disease. *Journal of Neural Transmission, 117*, 1019–1025.

Mangano, E. N., Litteljohn, D., So, R., Nelson, E., Peters, S., Bethune, C., et al. (2012). Interferon-gamma plays a role in paraquat-induced neurodegeneration involving oxidative and proinflammatory pathways. *Neurobiology of Aging, 33*, 1411–1426.

Manning-Bog, A. B., McCormack, A. L., Li, J., Uversky, V. N., Fink, A. L., & Di Monte, D. A. (2002). The herbicide paraquat causes up-regulation and aggregation of alpha-synuclein in mice: Paraquat and alpha-synuclein. *Journal of Biological Chemistry, 277*, 1641–1644.

Maragakis, N. J., & Rothstein, J. D. (2006). Mechanisms of disease: Astrocytes in neurodegenerative disease. *Nature Clinical Practice. Neurology, 2*, 679–689.

Marinova-Mutafchieva, L., Sadeghian, M., Broom, L., Davis, J. B., Medhurst, A. D., & Dexter, D. T. (2009). Relationship between microglial activation and dopaminergic neuronal loss in the substantia nigra: A time course study in a 6-hydroxydopamine model of Parkinson's disease. *Journal of Neurochemistry, 110*, 966–975.

Masliah, E., Rockenstein, E., Adame, A., Alford, M., Crews, L., Hashimoto, M., et al. (2005). Effects of alpha-synuclein immunization in a mouse model of Parkinson's disease. *Neuron, 46*, 857–868.
Masliah, E., Rockenstein, E., Mante, M., Crews, L., Spencer, B., Adame, A., et al. (2011). Passive immunization reduces behavioral and neuropathological deficits in an alpha-synuclein transgenic model of Lewy body disease. *PLoS One, 6*, e19338.
Masliah, E., Rockenstein, E., Veinbergs, I., Mallory, M., Hashimoto, M., Takeda, A., et al. (2000). Dopaminergic loss and inclusion body formation in alpha-synuclein mice: Implications for neurodegenerative disorders. *Science, 287*, 1265–1269.
Mattila, K. M., Rinne, J. O., Lehtimaki, T., Roytta, M., Ahonen, J. P., & Hurme, M. (2002). Association of an interleukin 1B gene polymorphism (-511) with Parkinson's disease in Finnish patients. *Journal of Medical Genetics, 39*, 400–402.
McCall, C. E., El Gazzar, M., Liu, T., Vachharajani, V., & Yoza, B. (2011). Epigenetics, bioenergetics, and microRNA coordinate gene-specific reprogramming during acute systemic inflammation. *Journal of Leukocyte Biology, 90*, 439–446.
McClain, J. A., Phillips, L. L., & Fillmore, H. L. (2009). Increased MMP-3 and CTGF expression during lipopolysaccharide-induced dopaminergic neurodegeneration. *Neuroscience Letters, 460*, 27–31.
McCoy, M. K., Ruhn, K. A., Blesch, A., & Tansey, M. G. (2011). TNF: A key neuroinflammatory mediator of neurotoxicity and neurodegeneration in models of Parkinson's disease. *Advances in Experimental Medicine and Biology, 691*, 539–540.
McGeer, P. L., Itagaki, S., Boyes, B. E., & McGeer, E. G. (1988). Reactive microglia are positive for HLA-DR in the substantia nigra of Parkinson's and Alzheimer's disease brains. *Neurology, 38*, 1285–1291.
McGeer, P. L., Schwab, C., Parent, A., & Doudet, D. (2003). Presence of reactive microglia in monkey substantia nigra years after 1-methyl-4-phenyl-1,2,3,6-tetrahydropyridine administration. *Annals of Neurology, 54*, 599–604.
McGeer, P. L., Yasojima, K., & McGeer, E. G. (2002). Association of interleukin-1 beta polymorphisms with idiopathic Parkinson's disease. *Neuroscience Letters, 326*, 67–69.
Melo, A., Monteiro, L., Lima, R. M., de Oliveira, D. M., de Cerqueira, M. D., & El-Bacha, R. S. (2011). Oxidative stress in neurodegenerative diseases: Mechanisms and therapeutic perspectives. *Oxidative Medicine and Cellular Longevity, 2011*, 467180.
Meredith, G. E., Totterdell, S., Potashkin, J. A., & Surmeier, D. J. (2008). Modeling PD pathogenesis in mice: Advantages of a chronic MPTP protocol. *Parkinsonism & Related Disorders, 14*(Suppl. 2), S112–S115.
Middeldorp, J., & Hol, E. M. (2011). GFAP in health and disease. *Progress in Neurobiology, 93*, 421–443.
Mirza, B., Hadberg, H., Thomsen, P., & Moos, T. (2000). The absence of reactive astrocytosis is indicative of a unique inflammatory process in Parkinson's disease. *Neuroscience, 95*, 425–432.
Mitra, S., Chakrabarti, N., & Bhattacharyya, A. (2011). Differential regional expression patterns of alpha-synuclein, TNF-alpha, and IL-1beta; and variable status of dopaminergic neurotoxicity in mouse brain after Paraquat treatment. *Journal of Neuroinflammation, 8*, 163.
Mittelbronn, M., Dietz, K., Schluesener, H. J., & Meyermann, R. (2001). Local distribution of microglia in the normal adult human central nervous system differs by up to one order of magnitude. *Acta Neuropathologica, 101*, 249–255.
Miyake, Y., Fukushima, W., Tanaka, K., Sasaki, S., Kiyohara, C., Tsuboi, Y., et al. (2011). Dietary intake of antioxidant vitamins and risk of Parkinson's disease: A case-control study in Japan. *European Journal of Neurology: The Official Journal of the European Federation of Neurological Societies, 18*, 106–113.
Mogi, M., Harada, M., Kondo, T., Riederer, P., Inagaki, H., Minami, M., et al. (1994a). Interleukin-1 beta, interleukin-6, epidermal growth factor and transforming growth

factor-alpha are elevated in the brain from parkinsonian patients. *Neuroscience Letters*, *180*, 147–150.

Mogi, M., Harada, M., Riederer, P., Narabayashi, H., Fujita, K., & Nagatsu, T. (1994b). Tumor necrosis factor-alpha (TNF-alpha) increases both in the brain and in the cerebrospinal fluid from parkinsonian patients. *Neuroscience Letters*, *165*, 208–210.

Mogi, M., Kondo, T., Mizuno, Y., & Nagatsu, T. (2007). p53 protein, interferon-gamma, and NF-kappaB levels are elevated in the parkinsonian brain. *Neuroscience Letters*, *414*, 94–97.

Moran, L. B., Croisier, E., Duke, D. C., Kalaitzakis, M. E., Roncaroli, F., Deprez, M., et al. (2007). Analysis of alpha-synuclein, dopamine and parkin pathways in neuropathologically confirmed parkinsonian nigra. *Acta Neuropathologica*, *113*, 253–263.

Moran, L. B., & Graeber, M. B. (2008). Towards a pathway definition of Parkinson's disease: A complex disorder with links to cancer, diabetes and inflammation. *Neurogenetics*, *9*, 1–13.

Muller, T., Blum-Degen, D., Przuntek, H., & Kuhn, W. (1998). Interleukin-6 levels in cerebrospinal fluid inversely correlate to severity of Parkinson's disease. *Acta Neurologica Scandinavica*, *98*, 142–144.

Munoz, A., Rey, P., Guerra, M. J., Mendez-Alvarez, E., Soto-Otero, R., & Labandeira-Garcia, J. L. (2006). Reduction of dopaminergic degeneration and oxidative stress by inhibition of angiotensin converting enzyme in a MPTP model of parkinsonism. *Neuropharmacology*, *51*, 112–120.

Muthane, U. B., Swamy, H. S., Satishchandra, P., Subhash, M. N., Rao, S., & Subbakrishna, D. (1994). Early onset Parkinson's disease: Are juvenile- and young-onset different? *Movement Disorders: Official Journal of the Movement Disorder Society*, *9*, 539–544.

Mythri, R. B., Venkateshappa, C., Harish, G., Mahadevan, A., Muthane, U. B., Yasha, T. C., et al. (2011). Evaluation of markers of oxidative stress, antioxidant function and astrocytic proliferation in the striatum and frontal cortex of Parkinson's disease brains. *Neurochemical Research*, *36*, 1452–1463.

Nalls, M. A., Plagnol, V., Hernandez, D. G., Sharma, M., Sheerin, U. M., Saad, M., et al. (2011). Imputation of sequence variants for identification of genetic risks for Parkinson's disease: A meta-analysis of genome-wide association studies. *Lancet*, *377*, 641–649.

Nicoletti, A., Fagone, P., Donzuso, G., Mangano, K., Dibilio, V., Caponnetto, S., et al. (2011). Parkinson's disease is associated with increased serum levels of macrophage migration inhibitory factor. *Cytokine*, *55*, 165–167.

Nimmerjahn, A. (2009). Astrocytes going live: Advances and challenges. *The Journal of Physiology*, *587*, 1639–1647.

NINSD NET-PD Investigators, (2008). A pilot clinical trial of creatine and minocycline in early Parkinson disease: 18-month results. *Clinical Neuropharmacology*, *31*, 141–150.

Nishimura, M., Kuno, S., Kaji, R., Yasuno, K., & Kawakami, H. (2005). Glutathione-S-transferase-1 and interleukin-1beta gene polymorphisms in Japanese patients with Parkinson's disease. *Movement Disorders: Official Journal of the Movement Disorder Society*, *20*, 901–902.

Nishimura, M., Mizuta, I., Mizuta, E., Yamasaki, S., Ohta, M., Kaji, R., et al. (2001). Tumor necrosis factor gene polymorphisms in patients with sporadic Parkinson's disease. *Neuroscience Letters*, *311*, 1–4.

Noda, M., Doi, Y., Liang, J., Kawanokuchi, J., Sonobe, Y., Takeuchi, H., et al. (2011). Fractalkine attenuates excito-neurotoxicity via microglial clearance of damaged neurons and antioxidant enzyme heme oxygenase-1 expression. *Journal of Biological Chemistry*, *286*, 2308–2319.

Norazit, A., Meedeniya, A. C., Nguyen, M. N., & Mackay-Sim, A. (2010). Progressive loss of dopaminergic neurons induced by unilateral rotenone infusion into the medial forebrain bundle. *Brain Research*, *1360*, 119–129.

Oberheim, N. A., Takano, T., Han, X., He, W., Lin, J. H., Wang, F., et al. (2009). Uniquely hominid features of adult human astrocytes. *The Journal of Neuroscience: The Official Journal of the Society for Neuroscience*, *29*, 3276–3287.

Ohnuki, T., Nakamura, A., Okuyama, S., & Nakamura, S. (2010). Gene expression profiling in progressively MPTP-lesioned macaques reveals molecular pathways associated with sporadic Parkinson's disease. *Brain Research*, *1346*, 26–42.

Okun, E., Griffioen, K. J., & Mattson, M. P. (2011). Toll-like receptor signaling in neural plasticity and disease. *Trends in Neurosciences*, *34*, 269–281.

Okun, E., Mattson, M. P., & Arumugam, T. V. (2010). Involvement of Fc receptors in disorders of the central nervous system. *Neuromolecular Medicine*, *12*, 164–178.

Orr, C. F., Rowe, D. B., Mizuno, Y., Mori, H., & Halliday, G. M. (2005). A possible role for humoral immunity in the pathogenesis of Parkinson's disease. *Brain: A Journal of Neurology*, *128*, 2665–2674.

Ouchi, Y., Yoshikawa, E., Sekine, Y., Futatsubashi, M., Kanno, T., Ogusu, T., et al. (2005). Microglial activation and dopamine terminal loss in early Parkinson's disease. *Annals of Neurology*, *57*, 168–175.

Pabon, M. M., Bachstetter, A. D., Hudson, C. E., Gemma, C., & Bickford, P. C. (2011). CX3CL1 reduces neurotoxicity and microglial activation in a rat model of Parkinson's disease. *Journal of Neuroinflammation*, *8*, 9.

Palacios, N., Gao, X., McCullough, M. L., Jacobs, E. J., Patel, A. V., Mayo, T., et al. (2011). Obesity, diabetes, and risk of Parkinson's disease. *Movement Disorders: Official Journal of the Movement Disorder Society*, *26*, 2253–2259.

Paolicelli, R. C., Bolasco, G., Pagani, F., Maggi, L., Scianni, M., Panzanelli, P., et al. (2011). Synaptic pruning by microglia is necessary for normal brain development. *Science*, *333*, 1456–1458.

Park, B., Yang, J., Yun, N., Choe, K. M., Jin, B. K., & Oh, Y. J. (2010). Proteomic analysis of expression and protein interactions in a 6-hydroxydopamine-induced rat brain lesion model. *Neurochemistry International*, *57*, 16–32.

Parkhurst, C. N., & Gan, W. B. (2010). Microglia dynamics and function in the CNS. *Current Opinion in Neurobiology*, *20*, 595–600.

Pattarini, R., Smeyne, R. J., & Morgan, J. I. (2007). Temporal mRNA profiles of inflammatory mediators in the murine 1-methyl-4-phenyl-1,2,3,6-tetrahydropyrimidine model of Parkinson's disease. *Neuroscience*, *145*, 654–668.

Peng, J., Xie, L., Stevenson, F. F., Melov, S., Di Monte, D. A., & Andersen, J. K. (2006). Nigrostriatal dopaminergic neurodegeneration in the weaver mouse is mediated via neuroinflammation and alleviated by minocycline administration. *The Journal of Neuroscience: The Official Journal of the Society for Neuroscience*, *26*, 11644–11651.

Perier, C., Bove, J., Wu, D. C., Dehay, B., Choi, D. K., Jackson-Lewis, V., et al. (2007). Two molecular pathways initiate mitochondria-dependent dopaminergic neurodegeneration in experimental Parkinson's disease. *Proceedings of the National Academy of Sciences of the United States of America*, *104*, 8161–8166.

Phani, S., Loike, J. D., & Przedborski, S. (2012). Neurodegeneration and inflammation in Parkinson's disease. *Parkinsonism & Related Disorders*, *18*(Suppl. 1), S207–S209.

Phinney, A. L., Andringa, G., Bol, J. G., Wolters, E., van Muiswinkel, F. L., van Dam, A. M., et al. (2006). Enhanced sensitivity of dopaminergic neurons to rotenone-induced toxicity with aging. *Parkinsonism & Related Disorders*, *12*, 228–238.

Pierre, S. R., Lemmens, M. A., & Figueiredo-Pereira, M. E. (2009). Subchronic infusion of the product of inflammation prostaglandin J2 models sporadic Parkinson's disease in mice. *Journal of Neuroinflammation*, *6*, 18.

Pourcher, E., Fernandez, H. H., Stacy, M., Mori, A., Ballerini, R., & Chaikin, P. (2011). Istradefylline for Parkinson's disease patients experiencing motor fluctuations: Results of the KW-6002-US-018 study. *Parkinsonism & Related Disorders*, *18*, 178–184.

Powers, K. M., Kay, D. M., Factor, S. A., Zabetian, C. P., Higgins, D. S., Samii, A., et al. (2008). Combined effects of smoking, coffee, and NSAIDs on Parkinson's disease risk. *Movement Disorders: Official Journal of the Movement Disorder Society, 23*, 88–95.

Prasad, K., Tarasewicz, E., Mathew, J., Strickland, P. A., Buckley, B., Richardson, J R., et al. (2009). Toxicokinetics and toxicodynamics of paraquat accumulation in mouse brain. *Experimental Neurology, 215*, 358–367.

Prediger, R. D., Batista, L. C., Medeiros, R., Pandolfo, P., Florio, J. C., & Takahashi, R. N. (2006). The risk is in the air: Intranasal administration of MPTP to rats reproducing clinical features of Parkinson's disease. *Experimental Neurology, 202*, 391–403.

Priyadarshi, A., Khuder, S. A., Schaub, E. A., & Priyadarshi, S. S. (2001). Environmental risk factors and Parkinson's disease: A metaanalysis. *Environmental Research, 86*, 122–127.

Przedborski, S. (2007). Neuroinflammation and Parkinson's disease. In P. J. Vinken & G. W. Bruyn (Eds.), *Handbook of clinical neurology83*, (pp. 535–551). New York: Springer.

Przedborski, S., Chen, Q., Vila, M., Giasson, B. I., Djaldatti, R., Vukosavic, S., et al. (2001). Oxidative post-translational modifications of alpha-synuclein in the 1-methyl-4-phenyl-1,2,3,6-tetrahydropyridine (MPTP) mouse model of Parkinson's disease. *Journal of Neurochemistry, 76*, 637–640.

Purisai, M. G., McCormack, A. L., Cumine, S., Li, J., Isla, M. Z., & Di Monte, D. A. (2007). Microglial activation as a priming event leading to paraquat-induced dopaminergic cell degeneration. *Neurobiology of Disease, 25*, 392–400.

Puschmann, A., Verbeeck, C., Heckman, M. G., Soto-Ortolaza, A. I., Lynch, T., Jasinska-Myga, B., et al. (2011). Human leukocyte antigen variation and Parkinson's disease. *Parkinsonism & Related Disorders, 17*, 376–378.

Qin, L., Liu, Y., Wang, T., Wei, S. J., Block, M. L., Wilson, B., et al. (2004). NADPH oxidase mediates lipopolysaccharide-induced neurotoxicity and proinflammatory gene expression in activated microglia. *Journal of Biological Chemistry, 279*, 1415–1421.

Qin, L., Wu, X., Block, M. L., Liu, Y., Breese, G. R., Hong, J. S., et al. (2007). Systemic LPS causes chronic neuroinflammation and progressive neurodegeneration. *Glia, 55*, 453–462.

Quinn, L. P., Crook, B., Hows, M. E., Vidgeon-Hart, M., Chapman, H., Upton, N., et al. (2008). The PPARgamma agonist pioglitazone is effective in the MPTP mouse model of Parkinson's disease through inhibition of monoamine oxidase B. *British Journal of Pharmacology, 154*, 226–233.

Quintero, E. M., Willis, L., Singleton, R., Harris, N., Huang, P., Bhat, N., et al. (2006). Behavioral and morphological effects of minocycline in the 6-hydroxydopamine rat model of Parkinson's disease. *Brain Research, 1093*, 198–207.

Ransohoff, R. M., & Cardona, A. E. (2010). The myeloid cells of the central nervous system parenchyma. *Nature, 468*, 253–262.

Ransohoff, R. M., & Perry, V. H. (2009). Microglial physiology: Unique stimuli, specialized responses. *Annual Review of Immunology, 27*, 119–145.

Ransohoff, R. M., & Stevens, B. (2011). Neuroscience. How many cell types does it take to wire a brain? *Science, 333*, 1391–1392.

Rappold, P. M., & Tieu, K. (2010). Astrocytes and therapeutics for Parkinson's disease. *Neurotherapeutics: The Journal of the American Society for Experimental NeuroTherapeutics*, 7, 413–423.

Rees, K., Stowe, R., Patel, S., Ives, N., Breen, K., Clarke, C. E., et al. (2011). Non-steroidal anti-inflammatory drugs as disease-modifying agents for Parkinson's disease: Evidence from observational studies. *Cochrane Database of Systematic Reviews*, CD008454.

Reksidler, A. B., Lima, M. M., Zanata, S. M., Machado, H. B., da Cunha, C., Andreatini, R., et al. (2007). The COX-2 inhibitor parecoxib produces neuroprotective effects in MPTP-lesioned rats. *European Journal of Pharmacology, 560*, 163–175.

Rentzos, M., Nikolaou, C., Andreadou, E., Paraskevas, G. P., Rombos, A., Zoga, M., et al. (2007). Circulating interleukin-15 and RANTES chemokine in Parkinson's disease. *Acta Neurologica Scandinavica, 116*, 374–379.

Rentzos, M., Nikolaou, C., Andreadou, E., Paraskevas, G. P., Rombos, A., Zoga, M., et al. (2009). Circulating interleukin-10 and interleukin-12 in Parkinson's disease. *Acta Neurologica Scandinavica, 119*, 332–337.

Rey, P., Lopez-Real, A., Sanchez-Iglesias, S., Munoz, A., Soto-Otero, R., & Labandeira-Garcia, J. L. (2007). Angiotensin type-1-receptor antagonists reduce 6-hydroxydopamine toxicity for dopaminergic neurons. *Neurobiology of Aging, 28*, 555–567.

Reynolds, A. D., Stone, D. K., Hutter, J. A., Benner, E. J., Mosley, R. L., & Gendelman, H. E. (2010). Regulatory T cells attenuate Th17 cell-mediated nigrostriatal dopaminergic neurodegeneration in a model of Parkinson's disease. *Journal of Immunology, 184*, 2261–2271.

Reynolds, A. D., Stone, D. K., Mosley, R. L., & Gendelman, H. E. (2009a). Nitrated {alpha}-synuclein-induced alterations in microglial immunity are regulated by CD4+ T cell subsets. *Journal of Immunology, 182*, 4137–4149.

Reynolds, A. D., Stone, D. K., Mosley, R. L., & Gendelman, H. E. (2009b). Proteomic studies of nitrated alpha-synuclein microglia regulation by CD4+CD25+ T cells. *Journal of Proteome Research, 8*, 3497–3511.

Rezai-Zadeh, K., Gate, D., & Town, T. (2009). CNS infiltration of peripheral immune cells: D-Day for neurodegenerative disease? *Journal of Neuroimmune Pharmacology: The Official Journal of the Society on NeuroImmune Pharmacology, 4*, 462–475.

Ricci, G., Volpi, L., Pasquali, L., Petrozzi, L., & Siciliano, G. (2009). Astrocyte-neuron interactions in neurological disorders. *Journal of Biological Physics, 35*, 317–336.

Rivest, S. (2009). Regulation of innate immune responses in the brain. *Nature Reviews. Immunology, 9*, 429–439.

Rockenstein, E., Mallory, M., Hashimoto, M., Song, D., Shults, C. W., Lang, I., et al. (2002). Differential neuropathological alterations in transgenic mice expressing alpha-synuclein from the platelet-derived growth factor and Thy-1 promoters. *Journal of Neuroscience Research, 68*, 568–578.

Rodnitzky, R. L. (2012). Upcoming treatments in Parkinson's disease, including gene therapy. *Parkinsonism & Related Disorders, 18*(Suppl. 1), S37–S40.

Roodveldt, C., Christodoulou, J., & Dobson, C. M. (2008). Immunological features of alpha-synuclein in Parkinson's disease. *Journal of Cellular and Molecular Medicine, 12*, 1820–1829.

Ros-Bernal, F., Hunot, S., Herrero, M. T., Parnadeau, S., Corvol, J. C., Lu, L., et al. (2011). Microglial glucocorticoid receptors play a pivotal role in regulating dopaminergic neurodegeneration in parkinsonism. *Proceedings of the National Academy of Sciences of the United States of America, 108*, 6632–6637.

Rose, S., Jackson, M. J., Smith, L. A., Stockwell, K., Johnson, L., Carminati, P., et al. (2006). The novel adenosine A2a receptor antagonist ST1535 potentiates the effects of a threshold dose of L-DOPA in MPTP treated common marmosets. *European Journal of Pharmacology, 546*, 82–87.

Rosi, S., Vazdarjanova, A., Ramirez-Amaya, V., Worley, P. F., Barnes, C. A., & Wenk, G. L. (2006). Memantine protects against LPS-induced neuroinflammation, restores behaviorally-induced gene expression and spatial learning in the rat. *Neuroscience, 142*, 1303–1315.

Ross, O. A., O'Neill, C., Rea, I. M., Lynch, T., Gosal, D., Wallace, A., et al. (2004). Functional promoter region polymorphism of the proinflammatory chemokine IL-8 gene associates with Parkinson's disease in the Irish. *Human Immunology, 65*, 340–346.

Rostene, W., Dansereau, M. A., Godefroy, D., Van Steenwinckel, J., Reaux-Le Goazigo, A., Melik-Parsadaniantz, S., et al. (2011). Neurochemokines: A menage a trois providing new insights on the functions of chemokines in the central nervous system. *Journal of Neurochemistry, 118*, 680–694.

Rota, E., Bellone, G., Rocca, P., Bergamasco, B., Emanuelli, G., & Ferrero, P. (2006). Increased intrathecal TGF-beta1, but not IL-12, IFN-gamma and IL-10 levels in Alzheimer's disease patients. *Neurological Sciences: Official Journal of the Italian Neurological Society and of the Italian Society of Clinical Neurophysiology, 27*, 33–39.

Saijo, K., & Glass, C. K. (2011). Microglial cell origin and phenotypes in health and disease. *Nature Reviews. Immunology, 11*, 775–787.

Saijo, K., Winner, B., Carson, C. T., Collier, J. G., Boyer, L., Rosenfeld, M. G., et al. (2009). A Nurr1/CoREST pathway in microglia and astrocytes protects dopaminergic neurons from inflammation-induced death. *Cell, 137*, 47–59.

Saiki, M., Baker, A., Williams-Gray, C. H., Foltynie, T., Goodman, R. S., Taylor, C. J., et al. (2010). Association of the human leucocyte antigen region with susceptibility to Parkinson's disease. *Journal of Neurology, Neurosurgery, and Psychiatry, 81*, 890–891.

Saint-Pierre, M., Tremblay, M. E., Sik, A., Gross, R. E., & Cicchetti, F. (2006). Temporal effects of paraquat/maneb on microglial activation and dopamine neuronal loss in older rats. *Journal of Neurochemistry, 98*, 760–772.

Salama, M., & Arias-Carrion, O. (2011). Natural toxins implicated in the development of Parkinson's disease. *Therapeutic Advances in Neurological Disorders, 4*, 361–373.

Salminen, A., Ojala, J., Kaarniranta, K., Haapasalo, A., Hiltunen, M., & Soininen, H. (2011). Astrocytes in the aging brain express characteristics of senescence-associated secretory phenotype. *The European Journal of Neuroscience, 34*, 3–11.

Samii, A., Etminan, M., Wiens, M. O., & Jafari, S. (2009). NSAID use and the risk of Parkinson's disease: Systematic review and meta-analysis of observational studies. *Drugs & Aging, 26*, 769–779.

Sanchez-Guajardo, V., Febbraro, F., Kirik, D., & Romero-Ramos, M. (2010). Microglia acquire distinct activation profiles depending on the degree of alpha-synuclein neuropathology in a rAAV based model of Parkinson's disease. *PLoS One, 5*, e8784.

Sarkar, C., Basu, B., Chakroborty, D., Dasgupta, P. S., & Basu, S. (2010). The immunoregulatory role of dopamine: An update. *Brain, Behavior, and Immunity, 24*, 525–528.

Scalzo, P., Kummer, A., Cardoso, F., & Teixeira, A. L. (2009). Increased serum levels of soluble tumor necrosis factor-alpha receptor-1 in patients with Parkinson's disease. *Journal of Neuroimmunology, 216*, 122–125.

Scalzo, P., Kummer, A., Cardoso, F., & Teixeira, A. L. (2010). Serum levels of interleukin-6 are elevated in patients with Parkinson's disease and correlate with physical performance. *Neuroscience Letters, 468*, 56–58.

Schapira, A. H., Cooper, J. M., Dexter, D., Clark, J. B., Jenner, P., & Marsden, C. D. (1990). Mitochondrial complex I deficiency in Parkinson's disease. *Journal of Neurochemistry, 54*, 823–827.

Schiess, M. C., Barnes, J. L., Ellmore, T. M., Poindexter, B. J., Dinh, K., & Bick, R. J. (2010). CSF from Parkinson disease patients differentially affects cultured microglia and astrocytes. *BMC Neuroscience, 11*, 151.

Schmid, C. D., Melchior, B., Masek, K., Puntambekar, S. S., Danielson, P. E., Lo, D. D., et al. (2009). Differential gene expression in LPS/IFNgamma activated microglia and macrophages: in vitro versus in vivo. *Journal of Neurochemistry, 109*(Suppl. 1), 117–125.

Schneeberger, A., Mandler, M., Mattner, F., & Schmidt, W. (2010). AFFITOME(R) technology in neurodegenerative diseases: The doubling advantage. *Human Vaccines, 6*, 948–952.

Schneeberger, A., Mandler, M., Mattner, F., & Schmidt, W. (2012). Vaccination for Parkinson's disease. *Parkinsonism & Related Disorders, 18*(Suppl. 1), S11–S13.

Schneider, J. S., Yuwiler, A., & Markham, C. H. (1986). Production of a Parkinson-like syndrome in the cat with N-methyl-4-phenyl-1,2,3,6-tetrahydropyridine (MPTP): Behavior, histology, and biochemistry. *Experimental Neurology, 91*, 293–307.

Schulte, T., Schols, L., Muller, T., Woitalla, D., Berger, K., & Kruger, R. (2002). Polymorphisms in the interleukin-1 alpha and beta genes and the risk for Parkinson's disease. *Neuroscience Letters*, *326*, 70–72.

Schwab, C., Klegeris, A., & McGeer, P. L. (2010). Inflammation in transgenic mouse models of neurodegenerative disorders. *Biochimica et Biophysica Acta*, *1802*, 889–902.

Searles Nielsen, S., Gallagher, L. G., Lundin, J. I., Longstreth, W. T., Jr., Smith-Weller, T., Franklin, G. M., et al. (2012). Environmental tobacco smoke and Parkinson's disease. *Movement Disorders: Official Journal of the Movement Disorder Society*, *27*, 293–296.

Serhan, C. N. (2011). The resolution of inflammation: The devil in the flask and in the details. *The FASEB Journal: Official Publication of the Federation of American Societies for Experimental Biology*, *25*, 1441–1448.

Sherer, T. B., Betarbet, R., Kim, J. H., & Greenamyre, J. T. (2003). Selective microglial activation in the rat rotenone model of Parkinson's disease. *Neuroscience Letters*, *341*, 87–90.

Shimoji, M., Pagan, F., Healton, E. B., & Mocchetti, I. (2009). CXCR4 and CXCL12 expression is increased in the nigro-striatal system of Parkinson's disease. *Neurotoxicity Research*, *16*, 318–328.

Sidoryk-Wegrzynowicz, M., Wegrzynowicz, M., Lee, E., Bowman, A. B., & Aschner, M. (2011). Role of astrocytes in brain function and disease. *Toxicologic Pathology*, *39*, 115–123.

Silverstein, F. E., Faich, G., Goldstein, J. L., Simon, L. S., Pincus, T., Whelton, A., et al. (2000). Gastrointestinal toxicity with celecoxib vs nonsteroidal anti-inflammatory drugs for osteoarthritis and rheumatoid arthritis: The CLASS study: A randomized controlled trial. Celecoxib Long-term Arthritis Safety Study. *JAMA: The Journal of the American Medical Association*, *284*, 1247–1255.

Simon-Sanchez, J., van Hilten, J. J., van de Warrenburg, B., Post, B., Berendse, H. W., Arepalli, S., et al. (2011). Genome-wide association study confirms extant PD risk loci among the Dutch. *European Journal of Human Genetics*, *19*, 655–661.

Simunovic, F., Yi, M., Wang, Y., Macey, L., Brown, L. T., Krichevsky, A. M., et al. (2009). Gene expression profiling of substantia nigra dopamine neurons: Further insights into Parkinson's disease pathology. *Brain: A Journal of Neurology*, *132*, 1795–1809.

Singh, S., Singh, K., Gupta, S. P., Patel, D. K., Singh, V. K., Singh, R. K., et al. (2009). Effect of caffeine on the expression of cytochrome P450 1A2, adenosine A2A receptor and dopamine transporter in control and 1-methyl 4-phenyl 1,2,3,6-tetrahydropyridine treated mouse striatum. *Brain Research*, *1283*, 115–126.

Sinha, A., Srivastava, N., Singh, S., Singh, A. K., Bhushan, S., Shukla, R., et al. (2009). Identification of differentially displayed proteins in cerebrospinal fluid of Parkinson's disease patients: A proteomic approach. *Clinica Chimica Acta: International Journal of Clinical Chemistry*, *400*, 14–20.

Smith, J. A., Das, A., Ray, S. K., & Banik, N. L. (2012). Role of pro-inflammatory cytokines released from microglia in neurodegenerative diseases. *Brain Research Bulletin*, *87*, 10–20.

Sofroniew, M. V., & Vinters, H. V. (2010). Astrocytes: Biology and pathology. *Acta Neuropathologica*, *119*, 7–35.

Spooren, A., Kolmus, K., Laureys, G., Clinckers, R., De Keyser, J., Haegeman, G., et al. (2011). Interleukin-6, a mental cytokine. *Brain Research Reviews*, *67*, 157–183.

Sriram, S. (2011). Role of glial cells in innate immunity and their role in CNS demyelination. *Journal of Neuroimmunology*, *239*, 13–20.

Stoessl, A. J., Martin, W. W., McKeown, M. J., & Sossi, V. (2011). Advances in imaging in Parkinson's disease. *Lancet Neurology*, *10*, 987–1001.

Stone, D. K., Reynolds, A. D., Mosley, R. L., & Gendelman, H. E. (2009). Innate and adaptive immunity for the pathobiology of Parkinson's disease. *Antioxidants & Redox Signaling*, *11*, 2151–2166.

Streit, W. J. (2006). Microglial senescence: Does the brain's immune system have an expiration date? *Trends in Neurosciences*, *29*, 506–510.

Streit, W. J., & Xue, Q. S. (2010). The Brain's aging immune system. *Aging and Disease*, *1*, 254–261.

Su, X., Maguire-Zeiss, K. A., Giuliano, R., Prifti, L., Venkatesh, K., & Federoff, H. J. (2008). Synuclein activates microglia in a model of Parkinson's disease. *Neurobiology of Aging*, *29*, 1690–1701.

Sugama, S., Wirz, S. A., Barr, A. M., Conti, B., Bartfai, T., & Shibasaki, T. (2004). Interleukin-18 null mice show diminished microglial activation and reduced dopaminergic neuron loss following acute 1-methyl-4-phenyl-1,2,3,6-tetrahydropyridine treatment. *Neuroscience*, *128*, 451–458.

Swanson, C. R., Joers, V., Bondarenko, V., Brunner, K., Simmons, H. A., Ziegler, T. E., et al. (2011). The PPAR-gamma agonist pioglitazone modulates inflammation and induces neuroprotection in parkinsonian monkeys. *Journal of Neuroinflammation*, *8*, 91.

Tajiri, N., Yasuhara, T., Shingo, T., Kondo, A., Yuan, W., Kadota, T., et al. (2010). Exercise exerts neuroprotective effects on Parkinson's disease model of rats. *Brain Research*, *1310*, 200–207.

Takano, T., Tian, G. F., Peng, W., Lou, N., Libionka, W., Han, X., et al. (2006). Astrocyte-mediated control of cerebral blood flow. *Nature Neuroscience*, *9*, 260–267.

Takeuchi, H. (2010). Neurotoxicity by microglia: Mechanisms and potential therapeutic strategy. *Clinical and Experimental Neuroimmunology*, *1*, 12–21.

Tanaka, K., Miyake, Y., Fukushima, W., Sasaki, S., Kiyohara, C., Tsuboi, Y., et al. (2010). Active and passive smoking and risk of Parkinson's disease. *Acta Neurologica Scandinavica*, *122*, 377–382.

Tansey, M. G., & Goldberg, M. S. (2010). Neuroinflammation in Parkinson's disease: Its role in neuronal death and implications for therapeutic intervention. *Neurobiology of Disease*, *37*, 510–518.

Tansey, M. G., McCoy, M. K., & Frank-Cannon, T. C. (2007). Neuroinflammatory mechanisms in Parkinson's disease: Potential environmental triggers, pathways, and targets for early therapeutic intervention. *Experimental Neurology*, *208*, 1–25.

Tapia-Gonzalez, S., Giraldez-Perez, R. M., Cuartero, M. I., Casarejos, M. J., Mena, M. A., Wang, X. F., et al. (2011). Dopamine and alpha-synuclein dysfunction in Smad3 null mice. *Molecular Neurodegeneration*, *6*, 72.

Teismann, P., & Ferger, B. (2001). Inhibition of the cyclooxygenase isoenzymes COX-1 and COX-2 provide neuroprotection in the MPTP-mouse model of Parkinson's disease. *Synapse*, *39*, 167–174.

Teismann, P., Tieu, K., Choi, D. K., Wu, D. C., Naini, A., Hunot, S., et al. (2003a). Cyclooxygenase-2 is instrumental in Parkinson's disease neurodegeneration. *Proceedings of the National Academy of Sciences of the United States of America*, *100*, 5473–5478.

Teismann, P., Vila, M., Choi, D. K., Tieu, K., Wu, D. C., Jackson-Lewis, V., et al. (2003b). COX-2 and neurodegeneration in Parkinson's disease. *Annals of the New York Academy of Sciences*, *991*, 272–277.

Theodore, S., Cao, S., McLean, P. J., & Standaert, D. G. (2008). Targeted overexpression of human alpha-synuclein triggers microglial activation and an adaptive immune response in a mouse model of Parkinson disease. *Journal of Neuropathology and Experimental Neurology*, *67*, 1149–1158.

Thevenet, J., Pescini Gobert, R., Hooft van Huijsduijnen, R., Wiessner, C., & Sagot, Y. J. (2011). Regulation of LRRK2 expression points to a functional role in human monocyte maturation. *PLoS One*, *6*, e21519.

Thomas, T., Timmer, M., Cesnulevicius, K., Hitti, E., Kotlyarov, A., & Gaestel, M. (2008). MAPKAP kinase 2-deficiency prevents neurons from cell death by reducing

neuroinflammation—Relevance in a mouse model of Parkinson's disease. *Journal of Neurochemistry, 105*, 2039–2052.
Tofaris, G. K., Garcia Reitbock, P., Humby, T., Lambourne, S. L., O'Connell, M., Ghetti, B., et al. (2006). Pathological changes in dopaminergic nerve cells of the substantia nigra and olfactory bulb in mice transgenic for truncated human alpha-synuclein(1-120): Implications for Lewy body disorders. *The Journal of Neuroscience: The Official Journal of the Society for Neuroscience, 26*, 3942–3950.
Tomas-Camardiel, M., Rite, I., Herrera, A. J., de Pablos, R. M., Cano, J., Machado, A., et al. (2004). Minocycline reduces the lipopolysaccharide-induced inflammatory reaction, peroxynitrite-mediated nitration of proteins, disruption of the blood-brain barrier, and damage in the nigral dopaminergic system. *Neurobiology of Disease, 16*, 190–201.
Ton, T. G., Heckbert, S. R., Longstreth, W. T., Jr., Rossing, M. A., Kukull, W. A., Franklin, G. M., et al. (2006). Nonsteroidal anti-inflammatory drugs and risk of Parkinson's disease. *Movement Disorders: Official Journal of the Movement Disorder Society, 21*, 964–969.
Tran, T. A., Nguyen, A. D., Chang, J., Goldberg, M. S., Lee, J. K., & Tansey, M. G. (2011). Lipopolysaccharide and tumor necrosis factor regulate Parkin expression via nuclear factor-kappa B. *PLoS One, 6*, e23660.
Tribl, G. G., Wober, C., Schonborn, V., Brucke, T., Deecke, L., & Panzer, S. (2001). Amantadine in Parkinson's disease: Lymphocyte subsets and IL-2 secreting T cell precursor frequencies. *Experimental Gerontology, 36*, 1761–1771.
Tufekci, K. U., Genc, S., & Genc, K. (2011). The endotoxin-induced neuroinflammation model of Parkinson's disease. *Parkinson's Disease, 2011*, 487450.
Valente, E. M., Arena, G., Torosantucci, L., & Gelmetti, V. (2012). Molecular pathways in sporadic PD. *Parkinsonism & Related Disorders, 18*(Suppl. 1), S71–S73.
van der Vliet, A., Eiserich, J. P., Halliwell, B., & Cross, C. E. (1997). Formation of reactive nitrogen species during peroxidase-catalyzed oxidation of nitrite. A potential additional mechanism of nitric oxide-dependent toxicity. *Journal of Biological Chemistry, 272*, 7617–7625.
Vargas, M. R., Johnson, D. A., Sirkis, D. W., Messing, A., & Johnson, J. A. (2008). Nrf2 activation in astrocytes protects against neurodegeneration in mouse models of familial amyotrophic lateral sclerosis. *The Journal of Neuroscience: The Official Journal of the Society for Neuroscience, 28*, 13574–13581.
Veerhuis, R., Nielsen, H. M., & Tenner, A. J. (2011). Complement in the brain. *Molecular Immunology, 48*, 1592–1603.
Veroni, C., Gabriele, L., Canini, I., Castiello, L., Coccia, E., Remoli, M. E., et al. (2010). Activation of TNF receptor 2 in microglia promotes induction of anti-inflammatory pathways. *Molecular and Cellular Neurosciences, 45*, 234–244.
von Bernhardi, R., Tichauer, J. E., & Eugenin, J. (2010). Aging-dependent changes of microglial cells and their relevance for neurodegenerative disorders. *Journal of Neurochemistry, 112*, 1099–1114.
Vroon, A., Drukarch, B., Bol, J. G., Cras, P., Breve, J. J., Allan, S. M., et al. (2007). Neuroinflammation in Parkinson's patients and MPTP-treated mice is not restricted to the nigrostriatal system: Microgliosis and differential expression of interleukin-1 receptors in the olfactory bulb. *Experimental Gerontology, 42*, 762–771.
Wachter, B., Schurger, S., Rolinger, J., von Ameln-Mayerhofer, A., Berg, D., Wagner, H. J., et al. (2010). Effect of 6-hydroxydopamine (6-OHDA) on proliferation of glial cells in the rat cortex and striatum: Evidence for de-differentiation of resident astrocytes. *Cell and Tissue Research, 342*, 147–160.
Wahner, A. D., Bronstein, J. M., Bordelon, Y. M., & Ritz, B. (2007). Nonsteroidal anti-inflammatory drugs may protect against Parkinson disease. *Neurology, 69*, 1836–1842.

Wandinger, K. P., Hagenah, J. M., Kluter, H., Rothermundt, M., Peters, M., & Vieregge, P. (1999). Effects of amantadine treatment on in vitro production of interleukin-2 in de-novo patients with idiopathic Parkinson's disease. *Journal of Neuroimmunology, 98*, 214–220.
Wang, J., Li, P. T., Du, H., Hou, J. C., Li, W. H., Pan, Y. S., et al. (2011a). Impact of paracrine signals from brain microvascular endothelial cells on microglial proliferation and migration. *Brain Research Bulletin, 86*, 53–59.
Wang, X. J., Zhang, S., Yan, Z. Q., Zhao, Y. X., Zhou, H. Y., Wang, Y., et al. (2011b). Impaired CD200-CD200R-mediated microglia silencing enhances midbrain dopaminergic neurodegeneration: Roles of aging, superoxide, NADPH oxidase, and p38 MAPK. *Free Radical Biology & Medicine, 50*, 1094–1106.
White, W. B. (2007). Cardiovascular effects of the cyclooxygenase inhibitors. *Hypertension, 49*, 408–418.
Whitehead, J. P. (2011). Diabetes: New conductors for the peroxisome proliferator-activated receptor gamma (PPARgamma) orchestra. *The International Journal of Biochemistry & Cell Biology, 43*, 1071–1074.
Willis, G. L., & Kennedy, G. A. (2004). The implementation of acute versus chronic animal models for treatment discovery in Parkinson's disease. *Reviews in the Neurosciences, 15*, 75–87.
Wilson, A. G., Symons, J. A., McDowell, T. L., McDevitt, H. O., & Duff, G. W. (1997). Effects of a polymorphism in the human tumor necrosis factor alpha promoter on transcriptional activation. *Proceedings of the National Academy of Sciences of the United States of America, 94*, 3195–3199.
Wirdefeldt, K., Adami, H. O., Cole, P., Trichopoulos, D., & Mandel, J. (2011). Epidemiology and etiology of Parkinson's disease: A review of the evidence. *European Journal of Epidemiology, 26*(Suppl. 1), S1–S58.
Witte, M. E., Geurts, J. J., de Vries, H. E., van der Valk, P., & van Horssen, J. (2010). Mitochondrial dysfunction: A potential link between neuroinflammation and neurodegeneration? *Mitochondrion, 10*, 411–418.
Wright, J. W., & Harding, J. W. (2011). Brain renin-angiotensin—A new look at an old system. *Progress in Neurobiology, 95*, 49–67.
Wu, H. M., Tzeng, N. S., Qian, L., Wei, S. J., Hu, X., Chen, S. H., et al. (2009). Novel neuroprotective mechanisms of memantine: Increase in neurotrophic factor release from astroglia and anti-inflammation by preventing microglial activation. *Neuropsychopharmacology: Official Publication of the American College of Neuropsychopharmacology, 34*, 2344–2357.
Wu, S. Y., Wang, T. F., Yu, L., Jen, C. J., Chuang, J. I., Wu, F. S., et al. (2011). Running exercise protects the substantia nigra dopaminergic neurons against inflammation-induced degeneration via the activation of BDNF signaling pathway. *Brain, Behavior, and Immunity, 25*, 135–146.
Xu, X., Li, D., He, Q., Gao, J., Chen, B., & Xie, A. (2011). Interleukin-18 promoter polymorphisms and risk of Parkinson's disease in a Han Chinese population. *Brain Research, 1381*, 90–94.
Xu, Q., Park, Y., Huang, X., Hollenbeck, A., Blair, A., Schatzkin, A., et al. (2010). Physical activities and future risk of Parkinson disease. *Neurology, 75*, 341–348.
Yadav, S., Gupta, S. P., Srivastava, G., Srivastava, P. K., & Singh, M. P. (2012). Role of secondary mediators in caffeine-mediated neuroprotection in maneb- and paraquat-induced Parkinson's disease phenotype in the mouse. *Neurochemical Research, 37*, 875–884.
Yang, M. S., Ji, K. A., Jeon, S. B., Jin, B. K., Kim, S. U., Jou, I., et al. (2006). Interleukin-13 enhances cyclooxygenase-2 expression in activated rat brain microglia: Implications for death of activated microglia. *Journal of Immunology, 177*, 1323–1329.
Yang, M. S., Min, K. J., & Joe, E. (2007). Multiple mechanisms that prevent excessive brain inflammation. *Journal of Neuroscience Research, 85*, 2298–2305.

Yasuda, Y., Shimoda, T., Uno, K., Tateishi, N., Furuya, S., Yagi, K., et al. (2008). The effects of MPTP on the activation of microglia/astrocytes and cytokine/chemokine levels in different mice strains. *Journal of Neuroimmunology*, *204*, 43–51.

Yin, G. N., Jeon, H., Lee, S., Lee, H. W., Cho, J. Y., & Suk, K. (2009). Role of soluble CD14 in cerebrospinal fluid as a regulator of glial functions. *Journal of Neuroscience Research*, *87*, 2578–2590.

Yirmiya, R., & Goshen, I. (2011). Immune modulation of learning, memory, neural plasticity and neurogenesis. *Brain, Behavior, and Immunity*, *25*, 181–213.

Yong, J., Lacan, G., Dang, H., Hsieh, T., Middleton, B., Wasserfall, C., et al. (2011). BCG vaccine-induced neuroprotection in a mouse model of Parkinson's disease. *PLoS One*, *6*, e16610.

Zecca, L., Wilms, H., Geick, S., Claasen, J. H., Brandenburg, L. O., Holzknecht, C., et al. (2008). Human neuromelanin induces neuroinflammation and neurodegeneration in the rat substantia nigra: Implications for Parkinson's disease. *Acta Neuropathologica*, *116*, 47–55.

Zecca, L., Zucca, F. A., Wilms, H., & Sulzer, D. (2003). Neuromelanin of the substantia nigra: A neuronal black hole with protective and toxic characteristics. *Trends in Neurosciences*, *26*, 578–580.

Zhang, W., Phillips, K., Wielgus, A. R., Liu, J., Albertini, A., Zucca, F. A., et al. (2011a). Neuromelanin activates microglia and induces degeneration of dopaminergic neurons: Implications for progression of Parkinson's disease. *Neurotoxicity Research*, *19*, 63–72.

Zhang, J., Sokal, I., Peskind, E. R., Quinn, J. F., Jankovic, J., Kenney, C., et al. (2008). CSF multianalyte profile distinguishes Alzheimer and Parkinson diseases. *American Journal of Clinical Pathology*, *129*, 526–529.

Zhang, J., Stanton, D. M., Nguyen, X. V., Liu, M., Zhang, Z., Gash, D., et al. (2005a). Intrapallidal lipopolysaccharide injection increases iron and ferritin levels in glia of the rat substantia nigra and induces locomotor deficits. *Neuroscience*, *135*, 829–838.

Zhang, W., Wang, T., Pei, Z., Miller, D. S., Wu, X., Block, M. L., et al. (2005b). Aggregated alpha-synuclein activates microglia: A process leading to disease progression in Parkinson's disease. *The FASEB Journal: Official Publication of the Federation of American Societies for Experimental Biology*, *19*, 533–542.

Zhang, S., Wang, X. J., Tian, L. P., Pan, J., Lu, G. Q., Zhang, Y. J., et al. (2011b). CD200-CD200R dysfunction exacerbates microglial activation and dopaminergic neurodegeneration in a rat model of Parkinson's disease. *Journal of Neuroinflammation*, *8*, 154.

Zhou, H. F., Liu, X. Y., Niu, D. B., Li, F. Q., He, Q. H., & Wang, X. M. (2005). Triptolide protects dopaminergic neurons from inflammation-mediated damage induced by lipopolysaccharide intranigral injection. *Neurobiology of Disease*, *18*, 441–449.

Zhou, Y. T., Yang, J. F., Zhang, Y. L., Wang, X. Y., & Chan, P. (2008). Protective role of interlekin-1 alpha gene polymorphism in Chinese Han population with sporadic Parkinson's disease. *Neuroscience Letters*, *445*, 23–25.

Zlokovic, B. V. (2008). The blood-brain barrier in health and chronic neurodegenerative disorders. *Neuron*, *57*, 178–201.

Zucker, I. H., & Ximmerman, M. C. (2011). The renin-angiotensin system in 2011: New avenues for translational research. *Current Opinion in Pharmacology*, *11*, 101–104.

CHAPTER FIVE

Treatment with Aβ42 Binding D-Amino Acid Peptides Reduce Amyloid Deposition and Inflammation in APP/PS1 Double Transgenic Mice

Thomas van Groen*†1, Inga Kadish*, Aileen Funke‡, Dirk Bartnik‡, Dieter Willbold‡§

*Department of Cell Biology, University of Alabama at Birmingham, Birmingham, Alabama, USA
†Department of Neurobiology, University of Alabama at Birmingham, Birmingham, Alabama, USA
‡Forschungszentrum Jülich, ICS-6, Jülich, Germany
§Institut für Physikalische Biologie and BMFZ, Heinrich-Heine-Universität Düsseldorf, Düsseldorf, Germany
[1]Corresponding author: e-mail address: vangroen@uab.edu

Contents

Abstract

One of the two characteristic pathological hallmarks of Alzheimer's disease (AD) are neuritic plaques. The sequence of events leading to the extracellular deposition of amyloid β (Aβ) peptides in plaques or in diffuse deposits is not clear. Here we investigate the relationship between aggregation and deposition of Aβ by using peptides that bind to Aβ as antifibrillization treatments in APP/PS1 double transgenic AD-model mice. Using Alzet minipumps, we infused the brain of these AD-model mice for 4 weeks with one of the three small D-amino acid peptides (i.e., D1, D3, or D3-FITC) that were designed to bind specifically to Aβ42, and examined the subsequent improvement in cognitive deficits after 3 weeks and analyzed amyloid deposition in the brain following the

Advances in Protein Chemistry and Structural Biology, Volume 88
ISSN 1876-1623
http://dx.doi.org/10.1016/B978-0-12-398314-5.00005-2

behavioral analysis. Cognitive deficits are similar comparing control and D3-treated mice, but D1-treated mice are slightly, but significantly, impaired. In contrast, there is a substantial improvement in the cognitive deficits in the animals treated with D3-FITC, compared to the other mice. In contrast, we show that there is a substantial reduction in the amount of amyloid deposits in the animals treated with D3, compared to the other groups of mice. Furthermore, the amount of activated microglia and astrocytes surrounding Aβ deposits is dramatically reduced in both the D3- and D3-FITC-treated mice. Our findings demonstrate that treatments with a high-affinity Aβ-42-binding D-amino acid peptide significantly decrease Aβ deposits and the associated inflammatory response. Together, this suggests that aggregation likely plays an important role in the deposition of Aβ protein in APP/PS1 transgenic mice and that antiaggregation treatments with D-peptides may be successful in AD patients.

1. INTRODUCTION

Alzheimer's disease (AD) is the most common form of dementia in the elderly (Hebert, Scherr, Bienias, Bennett, & Evans, 2003; Hirtz et al., 2007); the two characteristic pathological markers of the disease are significant numbers of neurofibrillary tangles (NFTs) and neuritic plaques in the brain (e.g., Hyman & Trojanowski, 1997). NFTs consist of hyperphosphorylated, twisted filaments of the cytoskeletal protein tau (e.g., Lee, Goedert, & Trojanowski, 2001), whereas plaques are primarily made up of amyloid β (Aβ; Dickson & Vickers, 2001; Selkoe, 2001), a 39–43-amino acid-long peptide derived from the proteolytic processing of the amyloid precursor protein (APP; e.g., Selkoe & Wolfe, 2007; Van Broeck, Van Broeckhoven, & Kumar-Singh, 2007). When APP is sequentially cleaved by the β-secretase and the γ-secretase, the resulting breakdown product is Aβ; in contrast, cleavage by α-secretase does not lead to Aβ production (Van Broeck et al., 2007).

Most cases of AD are sporadic; however, approximately 5% of AD cases are familial (Tanzi & Bertram, 2005), and these cases are related to mutations in the genes for APP and presenilin 1 and 2 (PS1 and PS2; Tanzi & Bertram, 2005). The mutations alter APP metabolism such that there is an increased production of Aβ (especially the longer, fibrillogenic Aβ42; Haass & De Strooper, 1999; Van Broeck et al., 2007). Together, this implies a central role for aberrant APP processing in the series of pathological changes occurring during AD, characterized by the appearance of the typical neuritic plaques (and NFTs) of AD (Braak & Braak, 1991, 1998).

Transgenic mice expressing mutated human AD genes offer a powerful model to study the role of Aβ in the development of AD pathology (e.g.,

German & Eisch, 2004). This study employs transgenic mice expressing both the human APPswe and PS1ΔE9 mutations (Jankowsky et al., 2001). These mice develop elevated levels of Aβ42 at about 4 months of age, and at around 5 months of age, they show amyloid β plaques (Jankowsky et al., 2001).

Several mechanisms have been proposed for the clearance of Aβ receptor-mediated Aβ transport across the blood–brain barrier and enzyme-mediated degradation of the peptide (Tanzi & Bertram, 2005; Wang, Zhou, & Zhou, 2006). Another mechanism that has been proposed is interference in the accumulation and deposition of the Aβ protein in aggregates (Masters & Beyreuther, 2006; Monaco, Zanusso, Mazzucco, & Rizzuto, 2006). We have developed a series of D-peptides, using mirror-image phage display techniques, that bind Aβ, with a preference for Aβ42 (Wiesehan & Willbold, 2003; Wiesehan et al., 2008). In this study, we used Alzet minipumps to infuse the hippocampus for 4 weeks with one of the three small D-amino acid peptides (i.e., D1, D3, or D3-FITC) that specifically bind Aβ42 (Wiesehan et al., 2003), and examined cognitive deficits after 3 weeks, and Aβ deposits in hippocampus and cortex following the final behavioral testing (i.e., after 4 weeks of treatment). The data demonstrate that infusion with the D3 peptide leads to a reduction in the size and number of Aβ deposits in the hippocampus and cortex, whereas the infusion with the D1 peptide or the vehicle solution does not change Aβ deposition patterns. It should be noted, however, that the 4-week infusion of either of the two peptides did not lead to a significant improvement in the cognitive deficits in these Tg AD-model mice. In contrast, treatment with D3-FITC significantly improved cognition, but did not substantially decrease amyloid deposition.

2. MATERIALS AND METHODS

2.1. Animals

APP and PS1 double transgenic mice (APPswe/PS1ΔE9 mice, $n=29$; Jankowsky et al., 2001) were used in this study. The mice were acquired from JAX at the age of 6 weeks, and until the treatments, the animals were housed 4/cage in our facility in a controlled environment (temperature 22 °C, humidity 50–60%, light from 07:00–19:00); food and water were available *ad libitum*. Following the implantation of the Alzet minipump, the animals were housed individually. The experiments were conducted according with the local Institutional Animal Care and Use Committee (IACUC) guidelines.

2.2. Peptides

A mirror-image phage display approach was used to identify novel and highly specific ligands for the AD amyloid peptide Aβ1–42 (Wiesehan et al., 2003). In short, a randomized 12-mer peptide library presented on M13 phages was screened for peptides with binding affinity for the mirror image of Aβ1–42 (Wiesehan et al., 2003, 2008). After four rounds of selection and amplification, the peptides were enriched with a dominating consensus sequence. The mirror image of the most representative peptide (i.e., D1; Table 5.1) was shown to bind Aβ1–42 with a dissociation constant in the submicromolar range (Wiesehan et al., 2003). For the other phage display selection procedure, the D-Aβ(1–42) was dissolved in HFIP (hexafluorisopropanol) in a concentration of 20 μM and diluted 1:10,000 in TBS (50 mM tris(hydroxymethyl)aminomethane (Tris)–HCl, pH 7.5, 150 mM NaCl). The final concentration of D-Aβ(1–42) was 2 nM. This solution was transferred into a streptavidin-coated tube (Roche-Boehringer, Mannheim, Germany) immediately and gently shaken for 15 min at room temperature. The tubes were washed two times with TBS and stored at −20 °C. For the selection procedure, 1×10^{11} phages (Ph.D.-12 Phage Display Peptide Library Kit; New England Biolabs, Frankfurt, Germany) were diluted in TBS-T with 0.1% BSA (w/w) and transferred into one of the prepared tubes. After 10-min incubation at room temperature, the tubes were washed 10 times with TBS-T and the elution of the binding phages was executed as described above. This selection procedure resulted in D3 (Table 5.1; Funke et al., 2010; Wiesehan et al., 2008).

2.3. Hippocampal infusion

In three groups of 8-month-old mice ($n=29$), we unilaterally infused the hippocampus for 4 weeks using Alzet osmotic minipumps, one group ($n=10$) with saline (peptides were dissolved in saline), one group with

Table 5.1 The nomenclature and amino acid sequence of the peptides used in this study

Peptide	Sequence	Description
D1	qshyrhispaqv	Dominating sequence selection 1, target: D-Aβ
D3	rprtrlhthrnr	Dominating sequence selection 3, target: D-Aβ

the D1 peptide ($n=9$), and one group with the D3 peptide ($n=10$). The peptide concentration was 0.25 mg/pump, that is, 0.25 mg in 250 μl, of the 0.25 mg peptide; 0.225 mg was unconjugated peptide; and 0.025 mg was peptide conjugated with an FITC molecule (to be able to visualize the infusion). The Alzet minipump (model #2004; delivery rate: 0.25 μl/h; duration: 4 weeks; Alzet) was soaked in sterile saline for 24 h; the next day, the pump, the connecting tube, and cannula (Alzet Brain Infusion Kit 3; Alzet) were filled with the appropriate solution, and they were connected, such that no air bubbles were present. Then the cannula was implanted in the brain (right dorsal hippocampus); in short, mice were anesthetized, placed in a stereotaxic frame, a hole was drilled above the right dorsal hippocampus (coordinates: A-2, L-1.5, V-2.2; Franklin & Paxinos, 1997), and the cannula was lowered into the hippocampus. Three and a half weeks after the start of the infusion, the animals were tested in the water maze; 5 weeks after the implantations, the animals were sacrificed for histopathological analysis (see below).

2.4. Behavior

The animals were tested for 1 week in an open-field water maze; our version of the maze consists of a blue circular tank of clear water (23 ± 1 °C). The mice were placed in the water at the edge of the pool and allowed to swim in order to locate a hidden, but fixed escape platform, using extramaze cues. On day 1, the mice were placed in the pool and allowed to swim freely for 90 s to find the hidden platform (or until they find the hidden platform); each animal is tested for four trials per day. A maximum swim time per trial of 90 s is allowed; if the animal does not locate the platform in that time, it is placed upon it by the experimenter and left there for 10 s. The intertrial interval is 60 s.

Each start position (east, north, south, and west) is used equally in a pseudorandom order, and the animals are always placed in the water facing the wall. All four possible quadrant positions for the platform locations are equally used among all of the animals. The platform is placed in the middle of one of the quadrants of the pool (i.e., northwest, southwest, northeast, or southeast; approximately 30 cm from the side of the pool). The mouse's task throughout the experiment is to find, and escape onto, the platform. Once the mouse has learned the task (day 5, trial 20), a probe trial is given immediately following the last trial of acquisition on day 5. In the probe trial (i.e., trial 21), the platform is removed from the pool and animals are allowed to swim for 60 s.

2.5. Histopathology

In short, mice were anesthetized, transcardially perfused, and the brains were removed. Following postfixation and cryoprotection, six series (1 in 6) of coronal sections were cut through the brain. The first series of sections was mounted unstained, the second, third, and fourth series were stained immunohistochemically according to published protocols (van Groen, Kiliaan, & Kadish, 2006), and the other two series were stored at -20 °C in antifreeze for future analysis. One half of the second series was stained for human Aβ using the W0-2 antibody (mouse anti-human $A\beta_{4-9}$; Ida et al., 1996), and the other half of the second series was stained for mouse Aβ (rabbit anti-rodent Aβ, Covance; van Groen et al., 2006). The first half of the third series was immunohistochemically stained for Aβ40 (mouse anti-Aβ40, Covance) and the other half for Aβ42 (mouse anti-Aβ42, Covance). One half of the fourth series was stained for GFAP (mouse anti-GFAP; Sigma), whereas the other half was stained for CD11b (rat anti-mouse CD11b; Serotec), a marker of microglia. Some of these stained sections were double stained with Congo red, thioflavin S, or thiazine red. The sections destined for Aβ staining were pretreated for 30 min with hot (85 °C) citrate buffer. The series of sections were transferred to a solution containing the primary antibody; this solution consists of TBS with 0.5% Triton X-100 added (TBS-T). Following incubation in this solution for 18 h on a shaker table at room temperature (20 °C) in the dark, the sections were rinsed three times in TBS-T and transferred to the solution containing the secondary antibody (goat anti-mouse★biotin; Sigma or sheep anti-rat Ig★biotin; Serotec). After 2 h, the sections were rinsed three times with TBS-T and transferred to a solution containing mouse ExtrAvidin® (Sigma); following rinsing, the sections were incubated for approximately 3 min with Ni-enhanced DAB (Kadish & van Groen, 2002). In a small number of sections, the Aβ deposits were double labeled for Aβ40, Aβ42, GFAP, or CD11b using fluorescent secondary antibodies. All stained sections were mounted on slides and coverslipped.

3. QUANTIFICATION

The appropriate areas (dorsal hippocampus and frontal cortex) of the brain were digitized using an Olympus DP70 digital camera, and the images were converted to gray scale using the Paint Shop Pro 7 program (Kadish & van Groen, 2002). To avoid changes in lighting, which might affect measurements, all images were acquired in one session. Further, to avoid differences in staining density between sections, the measurements were performed on sections that were stained simultaneously, that is, in the same staining tray ($n=24$). The percentage of area covered by the reaction product to Aβ was measured (van Groen et al., 2006) in the ipsi- and contralateral hippocampus (and ipsi- and contralateral frontal cortex) using the ScionImage (NIH) program (Kadish & van Groen, 2002). Employing a similar procedure, using digital images to be able to overlay the defined measurement area, plaques were counted in the same brain area on the adjacent sections that were stained with Congo red. The density of GFAP or CD11b staining was measured by placing a standard sized circle (200 μm diameter) around the plaque core (stained with Congo red) and measuring the optical density of the staining in the circle using the ScionImage (NIH) program. All density measurements were done in triplicate, that is, measuring the standardized area around three plaques at three different levels of the dorsal hippocampus and the frontal, midline cortex. These measurements were done in triplicate in sections (that had been stained simultaneously) by an observer blinded to the treatment of the animal (Kadish & van Groen, 2002). Data were analyzed by ANOVA (Systat 11; between groups), and post hoc tests (Tukey and Scheffe) were carried out to determine the source of a significant main effect or interaction.

4. RESULTS

The implantation of the Alzet minipumps did not change any obvious physiological parameters (e.g., growth as measured by body weight (Table 5.2) or general health, that is, look of fur, posture, and motor activity) in the implanted mice or cause any noticeable discomfort. None of the animals lost the brain cannula or developed any other health or motor problems.

After 3 weeks of control or D-peptide infusion, the three groups (i.e., control, D1, and D3) of APP-PSΔ mice were tested in the water maze. There was no significant difference in the swimming speed between the groups of mice (17.0 ± 3.5, 16.2 ± 3.3, and 16.2 ± 2.1 m/s, respectively). At the end of the week of training, there was a significant difference in escape latency between the three groups of mice in the water maze performance, but not in the probe trial performance (Fig. 5.1). Both the control and D3 groups of mice showed a small but significant improvement of learning during the week of testing, but did not show a difference between themselves; however, the D1 mice did not improve their performance during the week

Table 5.2 The number of animals per group, body weight, the Aβ load in the hippocampus (HIP) and frontal, midline cortex (FC), the number of Congo red (CR) positive plaques, and the density of GFAP and CD11b staining around plaques in the dorsal hippocampus are indicated

Group	**Control**	**D1**	**D3**
Infusion	Saline	D1 peptide	D3 peptide
Number, *n*	10	9	10
Body weight (g)	34.7±1.4	35.1±3.5	34.7±1.8
Aβ load (HIP) (%)	2.1±0.4	2.1±0.2	1.4±0.1*
Aβ load (FC) (%)	2.2±0.6	2.8±0.4	1.5±0.2*
CR# (HIP)	31±5	37±5	22±4*
CR# (FC)	49±9	51±8	31±6*
GFAP	102±2	101±2	89±3*
CD11b	159±3	165±3	147±2*

*Significantly different from control and D1, $p < 0.05$.

of training. Further, none of the groups showed a preference for the correct quadrant in the probe trial which followed the last training trial (Fig. 5.1). After completion of the behavioral testing, the animals were sacrificed and the brains were assessed for AD pathology.

The implantation of the cannula was in the dorsal hippocampus in all animals (Fig. 5.2). The extent of the distribution of the infused peptide in the brain was analyzed in the unstained sections of the brain since part (i.e., 10%) of the infused D-peptide was conjugated to an FITC molecule. Inspection of brain sections with an epifluorescent microscope revealed that all dense Aβ deposits in the whole brain were labeled, with a decrease in brightness further from the infusion site (not illustrated). Additionally, the deposits in the D3-infused mice were brighter than similarly located deposits (i.e., at the same distance from the infusion site) in the D1-infused mice (Fig. 5.2). It should be noted that, whereas all plaques, that is, Aβ deposits with a Congo red positive core, were labeled, neither diffuse Aβ deposits nor Aβ deposits in blood vessel walls were labeled, and none of the control-infused mice showed labeled Aβ deposits (Fig. 5.2). While the D1 peptide was only present at Aβ deposits, a small amount of the D3 peptide had also been taken up by neurons near the infusion site (Fig. 5.2). On the other hand, it is notable that neither near the infusion site nor anywhere else in

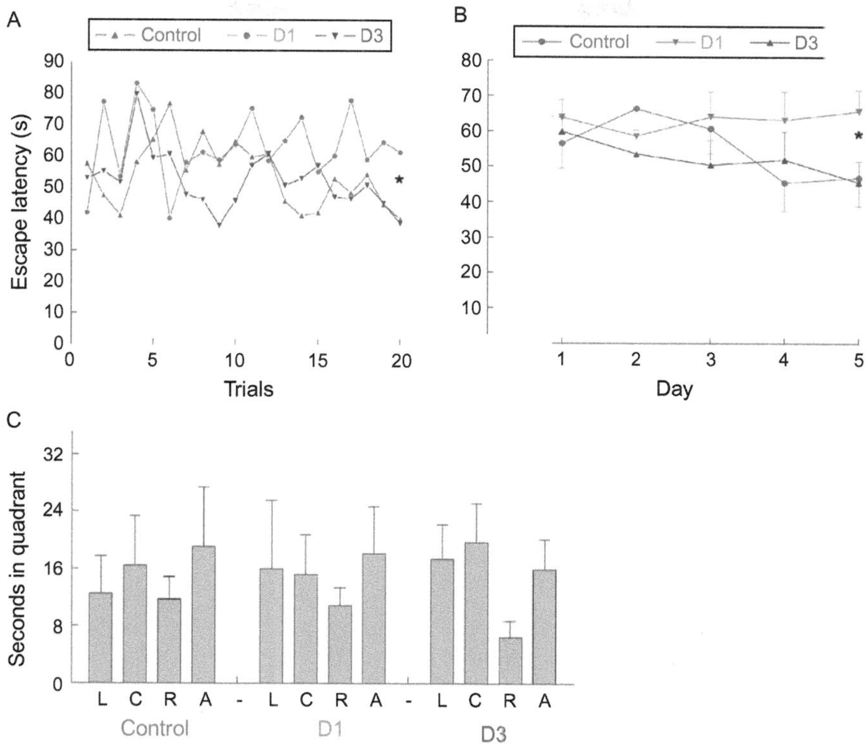

Figure 5.1 Three graphs showing the learning curves of the three groups of mice: (A) the per trial performance, and (B) the daily mean performance in the water maze, and (C) the probe trial performance of the three groups of mice. L, left; C, correct; A, across; and R, right quadrant. (For color version of this figure, the reader is referred to the online version of this chapter.)

the brain did any glial cells (i.e., astrocytes, microglia, or oligodendrocytes) show any signs of uptake of either of the two D-peptides.

Whereas labeled neurons were present near the infusion site of the D3-infused mice, in the D3 infused mice, a small number of layer II neurons in the entorhinal cortex also showed intracellular labeling with the D3 peptide (Fig. 5.3), likely due to D3 that was retrogradely transported from the infusion site, that is, the dorsal hippocampus. It should be noted that the labeled neurons in the entorhinal cortex were only present ipsilateral to the infusion; no labeled neurons were present contralateral to the infusion side. Further, while D3-labeled plaques in the entorhinal cortex showed plaque-related inflammation, the D3-labeled neurons did not show any signs of inflammation (Fig. 5.3). The D1-infused brains showed no uptake of the peptide near the infusion site or showed any transport of the peptide.

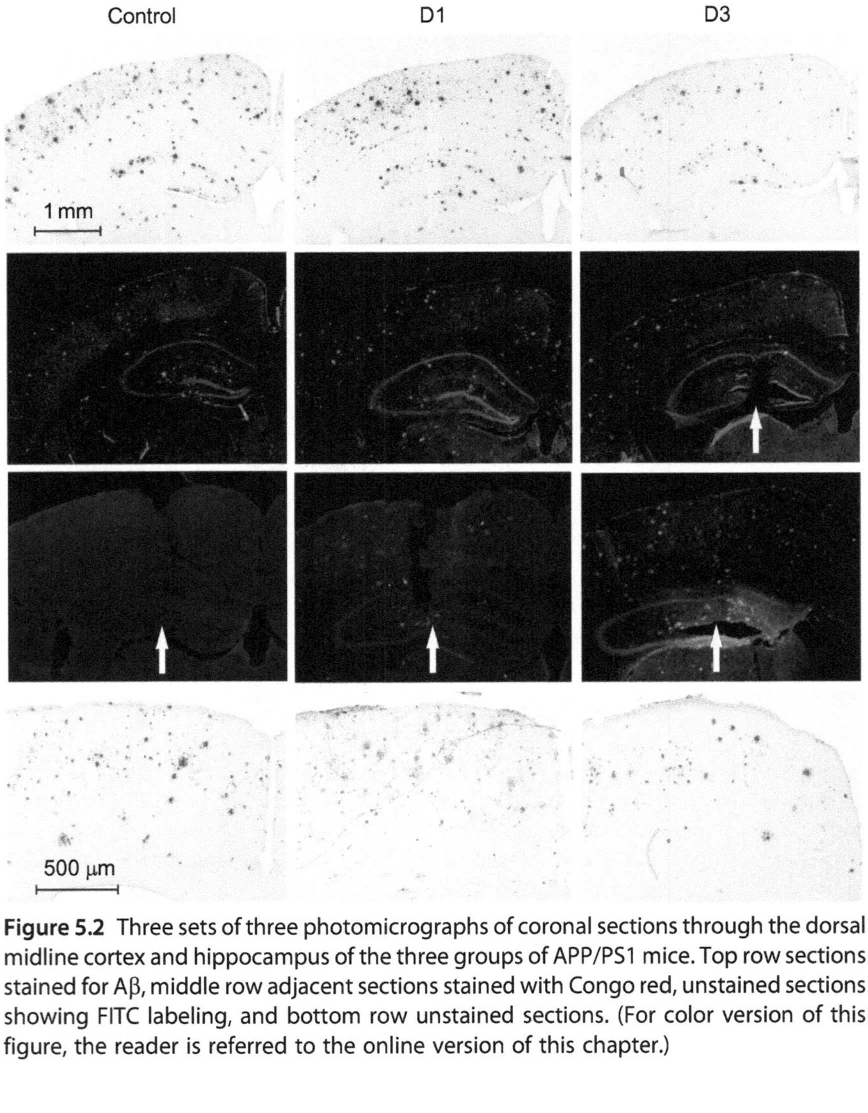

Figure 5.2 Three sets of three photomicrographs of coronal sections through the dorsal midline cortex and hippocampus of the three groups of APP/PS1 mice. Top row sections stained for Aβ, middle row adjacent sections stained with Congo red, unstained sections showing FITC labeling, and bottom row unstained sections. (For color version of this figure, the reader is referred to the online version of this chapter.)

In the sections that were stained for human $A\beta_{4-10}$ (using the W0-2 antibody), the Aβ load was measured in the hippocampus and frontal cortex. In the group that was infused with the D3 peptide, there was significant reduction in the Aβ load in the hippocampus and in the frontal cortex, both ipsilateral and contralateral to the infusion compared to the control-infused and the D1-infused mice (Fig. 5.2; Table 5.2). In contrast, there was no difference in the Aβ load between the control and the D1-infused mice in any brain area (Fig. 5.2; Table 5.2). The analysis of labeling for mouse

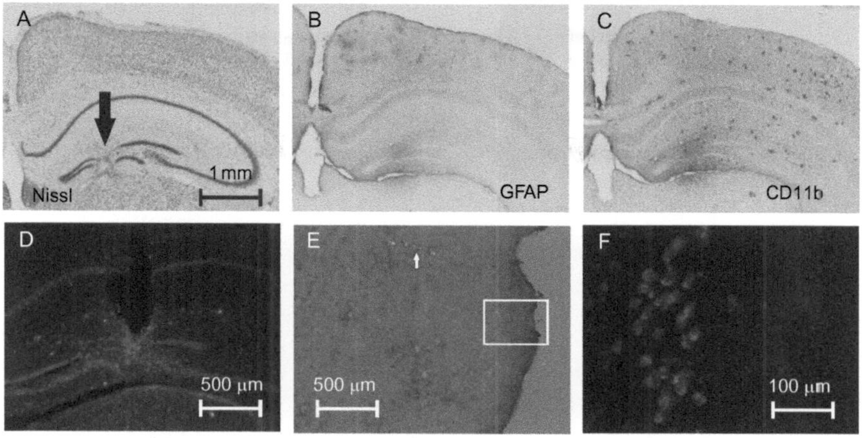

Figure 5.3 Photomicrographs of coronal sections through the brain of a mouse that was infused with D3 showing the infusion site. (A) Nissl stain, (B) adjacent GFAP-stained section, (C) adjacent microglia- (CD11b) stained section, and (D) unstained section showing infusion site with FITC labeling. (E) Entorhinal cortex of a D3-infused mouse, with combined GFAP, Congo red, and D3 staining, and (F) is the high power image of the box in (E), demonstrating retrogradely D3-FITC-labeled neurons in layer II of the entorhinal cortex. (For color version of this figure, the reader is referred to the online version of this chapter.)

Aβ showed a small but significant decrease in mouse Aβ load between the D3 group and the other groups. Furthermore, immunohistochemical staining for Aβ40 and Aβ42 showed, similar to the $A\beta_{4-10}$ staining, a decrease in the amount of staining for Aβ40 and Aβ42 in the D3-treated mice compared to the control and D1 groups.

Analysis of the sections that were stained for GFAP or microglia revealed that the implantation of the cannula and the infusions did not cause any significant inflammation or obvious pathology (Fig. 5.3). The only infusion-related inflammation that was present was around the infusion cannula (Fig. 5.3) and was similar among the three groups of mice. A more detailed analysis of the magnitude of inflammation surrounding Aβ deposits in the frontal, midline cortex (and other cortical areas), however, revealed that the D3 treatment significantly reduced the amount of plaque-related inflammation (measured in both GFAP and CD11b staining) compared to the control and D1-infused brains (Fig. 5.4; Table 5.2). Both astrocytes and microglial cells were significantly less activated at plaques in the D3-infused animals (Figs. 5.3 and 5.4). Double staining for microglia and GFAP confirmed that at D3-labeled plaques, both glial cell types were significantly less activated (Fig. 5.5).

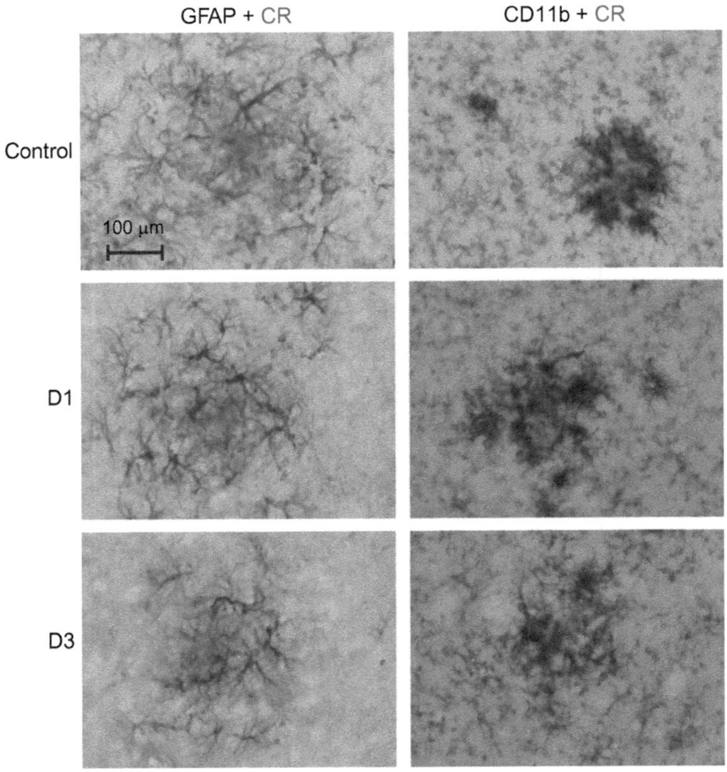

Figure 5.4 Photomicrographs demonstrating double-labeled (i.e., glial markers and Congo red) coronal sections of the midline cortex of mice that were control (top), D1 (middle), or D3 (bottom) infused. Left column stained for GFAP and Congo red (CR) and right column stained for microglia (CD11b) and Congo red. Note the reduced inflammation surrounding plaques in the D3-infused mouse brains. (For color version of this figure, the reader is referred to the online version of this chapter.)

Neither the D1 peptide nor the D3 peptide labeled any diffuse Aβ deposits or blood vessel Aβ deposits (not illustrated), but pericytes/perivascular macrophages throughout the brain showed labeling with D3 (Figs. 5.5 and 5.6). It should be noted that these were mainly present on penetrating arterioles (Fig. 5.6).

5. DISCUSSION

In this study, we infused the hippocampus of Tg AD-model mice with one of the two small D-amino acid peptides (i.e., either D1 or D3) that have a high affinity for Aβ42 (Wiesehan et al., 2003, 2008) and examined cognitive

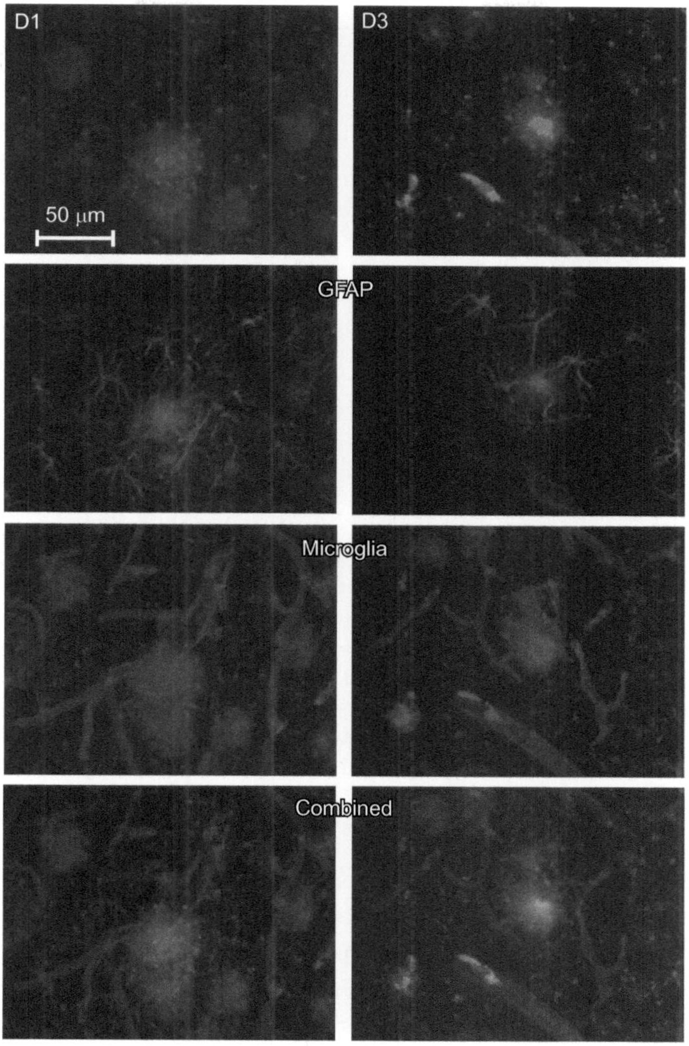

Figure 5.5 High power photomicrographs demonstrating triple-stained coronal sections of the midline cortex of mice that were infused with D1 (left) or D3 (right). Top row, D-peptides; second row, GFAP (blue); third row, microglia (red, tomato lectin); bottom row, combined image. Note the reduced inflammation surrounding plaques in the D3-infused mouse brains. (For interpretation of the references to color in this figure legend, the reader is referred to the online version of this chapter.)

performance after 3 weeks, and the Aβ deposits in hippocampus and cortex 1 weeks later, that is, after 4 weeks. There was no difference in cognitive deficits between control and D3-treated mice, but D1-treated mice were slightly more impaired compared to the other two groups. The data

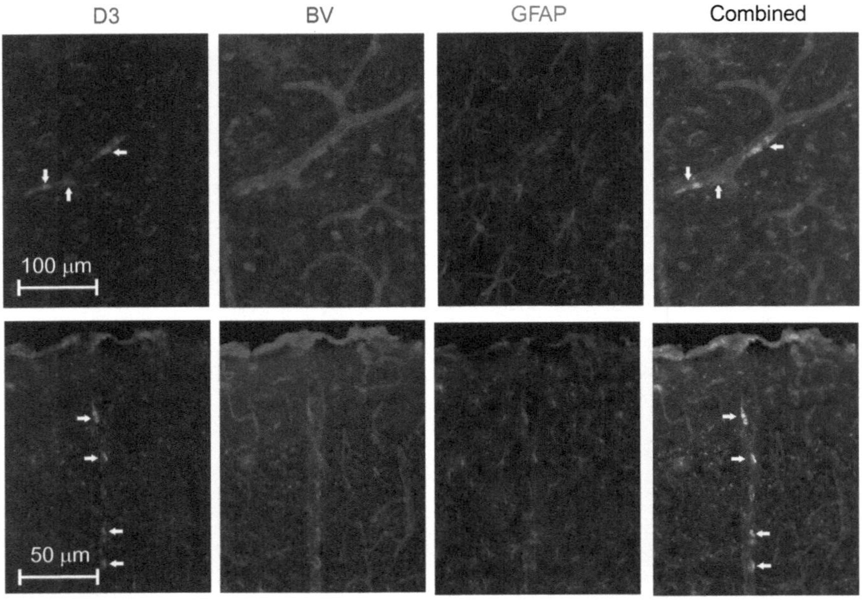

Figure 5.6 Photomicrographs demonstrating triple-stained coronal sections of the midline cortex of mice that were D3 infused, demonstrating D3-labeled perivascular macrophages. Left, D3-FITC labeling; red, blood vessel labeling with Texas Red-conjugated Tomato lectin; blue, GFAP staining; and last column, combined image. Arrows indicate D3-FITC-labeled perivascular macrophages. (For interpretation of the references to color in this figure legend, the reader is referred to the online version of this chapter.)

demonstrate that short-term infusion with the D3 peptide leads to a reduction in the density of Aβ deposits in the hippocampus and cortex, whereas the infusion with the D1 peptide or the vehicle solution does not change Aβ deposition patterns. Furthermore, analysis of glial stainings (i.e., GFAP and CD11b) shows that in the D3-infused mice, there is a significant reduction in both activated astrocytes and microglial cells surrounding Aβ plaques compared to the control and D1-treated mice.

The performance of the three groups of mice in the water maze is quite similar to our (and others) earlier reports (Liu et al., 2002; Malm et al., 2007), that is, these Tg AD-model mice are significantly impaired in their water maze performance at this age (9 months of age). At the end of the week of testing, both control and D3 groups of animals have slightly, but significantly, improved their performance; in contrast, the D1-infused mice have not improved their performance. This may possibly be related to the slight increase in plaque number in this group of mice compared to the other groups. Another possibility is that the binding of D1 to Aβ42 increases synaptic pathology, even without significantly changing Aβ load.

One of the two characteristic pathological hallmarks of human AD is the presence of neuritic plaques (Braak & Braak, 1991). Neuritic plaques have a dense core of aggregated Aβ peptides (e.g., Selkoe, 2001). However, diffuse amyloid deposits have also been shown to be present in AD brains (Braak & Braak, 1998; Dickson & Vickers, 2001), and they have been shown to be present in AD-model mice (van Groen et al., 2006). In our experiments, all plaques are labeled by both the D1 and the D3 peptide, and the data indicate that D3 has a higher affinity for Aβ42 in plaques than D1, since they are more brightly fluorescent (van Groen, Kadish, Wiesehan, Funke, & Willbold, 2009; Wiesehan et al., 2008). In contrast, diffuse deposits, that is, Aβ deposits without a thioflavin S positive core, do not show any labeling with either peptide; neither Aβ in blood vessel walls (CAA) is labeled by the peptides. The diffuse deposits do not show much, if any, labeling with Aβ42 (or Aβ40) specific antibodies (van Groen et al., 2006), which is likely why neither of the D-peptides binds to these deposits. It should be noted that while thioflavin S does stain CAA, the D-peptides do not label these deposits; however, the Aβ deposits in blood vessel walls consist predominantly of Aβ40 (Kumar-Singh, 2008) and are therefore not labeled by the D-peptides. In earlier *in vitro* experiments, at high ages (15 months of age) when blood vessel wall deposits also contain Aβ42, then these deposits are labeled by D1 and D3 (van Groen et al., 2009).

The data from our experiments show that the D3-infused mice have a lower Aβ load, demonstrating that binding of D3 to Aβ42 reduces Aβ deposition in the hippocampus and neocortex. This, most likely, is brought about by changes in the aggregation properties of the Aβ42–D3 complex, compared to the high property of Aβ42 to aggregate (Bartolini et al., 2007; Iijima et al., 2004). This would then lead to a decrease in the Aβ deposition rate, and thus a lesser number, and/or smaller plaques (Wiesehan et al., 2008). Several studies have shown that the use of β-sheet breakers (Permanne, Adessi, Fraga, et al., 2002; Permanne, Adessi, Saborio, et al., 2002; Soto, Kindy, Baumann, & Frangione, 1996) can reduce amyloid β aggregation and amyloid β deposition, both *in vitro* and *in vivo*. Our data confirm that even treatment with an Aβ42-binding peptide also will decrease amyloid deposition. On the other hand, it is possible that the clearance of the D3–Aβ42 complex has improved, compared to the clearance of Aβ42 (Ji, Permanne, Sigurdsson, Holtzman, & Wisniewski, 2001; Tanzi, Moir, & Wagner, 2004; Wang et al., 2006).

Several mechanisms have been proposed for the clearance of Aβ receptor-mediated Aβ transport across the blood–brain barrier and enzyme-mediated degradation of the peptide (Ji et al., 2001; Wang et al., 2006). It is more likely that Aβ transport out of the brain is improved than that Aβ degradation has increased, since the properties of the D3–Aβ42 complex are likely not improving enzymatic degradation (Yoshiike, Chui, Akagi, Tanakama, & Takashima, 2003).

Our data show that treatment with an Aβ42-binding peptide (D3) decreases amyloid deposition. It is of interest to note that plaque load—as measured in human Aβ-stained tissue—is significantly reduced in these mice compared to control or D1-treated mice. Similarly, in the D3 mice, plaque load is reduced in rodent Aβ-stained material and in material stained for c-terminal Aβ, that is, Aβ40 and Aβ42. The treatment with the Aβ42-binding D-peptides, however, does not seem to interfere with the staining for Aβ42.

It should be noted that all plaques in our AD-model mice are accompanied by activated glial cells, both astrocytes and microglia (e.g., Bondolfi et al., 2002; van Groen & Kadish, 2005; van Groen et al., 2011). The role of the activated glial cells is unclear: on the one hand, they could protect the brain by removing Aβ, whereas on the other hand, they can secrete inflammatory cytokines and generate NO and can thus damage and kill bystander neurons (Akiyama et al., 2000). It should be noted that in our mouse model, no dead or dying neurons are present, even near plaques (Wang, Tanila, Puolivali, Kadish, & van Groen, 2003). The role of activated microglia cells in the uptake of Aβ is disputed, with some groups showing clearance of Aβ by microglia (e.g., Rogers & Lue, 2001), whereas others have shown that microglia do not take up Aβ (Stalder, Deller, Staufenbiel, & Jucker, 2001). Similarly, the plaques in all our animals are surrounded by activated glial cells; however, neither in the D1- or the D3-infused mice, any fluorescence is present in microglial cells. In the D3-infused mice, compared to the D1-infused mice, however, the dense Aβ deposits are associated with much lower numbers of activated microglia or activated astrocytes. This suggests that the binding of D3 to deposited Aβ42 changes the structure of Aβ42 evoking less inflammatory responses (Iijima et al., 2004; Yoshiike et al., 2003). Furthermore, the size of the Aβ deposits is smaller in the D3-treated mice; together, this suggests that the Aβ in the plaques, through binding with the D3 peptide, has lower aggregation kinetics. Finally, it should be noted that the number of deposits that stain for thioflavin S or Congo red has also been significantly reduced. This indicates that either less new deposits were formed during the 4 weeks of treatment or deposits were removed by microglia. It is not likely that microglial cells have taken up Aβ42–D3–FITC, since no intracellular fluorescence is observed in any microglial cells. On the other hand, clear FITC fluorescence is present in pericytes/perivascular macrophages (Pluta et al., 2000) indicating that these cells have taken up the D3–FITC–Aβ42 complex. These cells may have increased the clearance of the complex from the brain through the blood–brain barrier (Pluta et al., 2000; Tanzi et al., 2004).

It has been suggested that the progression of the pathology in AD is related to the connections between areas displaying early deposits and the cortical regions showing later pathology, summarized by Hardy (1992) in the "anatomical cascade hypothesis." Our data showing labeled neurons in layer II of the entorhinal cortex that are most likely due to the retrograde transport of D3-FITC (or a D3–FITC–Aβ42 complex) confirm that Aβ is transported through axons (Buxbaum et al., 1998; Sheng, Price, & Koliatsos, 2002; van Groen, Liu, Ikonen, & Kadish, 2003), both in the anterograde and retrograde direction. It should be noted that the neurons showing intracellular D3 are healthy and do not show signs of cell death; furthermore, there is no increased inflammation in the area of labeled neurons. The high number of positive charges at physiological pH would make passive diffusion through the cell membrane very unlikely for D3. Adsorptive-mediated transcytosis has been proposed for the transport of several arginine-rich peptides, such as basic peptides derived from human immunodeficiency virus type 1 (HIV-1) Tat proteins, across the BBB (Drin, Cottin, Blanc, Rees, & Temsamani, 2003). Surprisingly, compared to D3, D1 shows lower BBB permeability than predicted from its lipophilicity (Liu, Funke, & Willbold, 2010). A likely possibility is that D3-FITC is actively taken up into the cell when it is bound to the Aβ42 peptide, and since D1-FITC binds to different forms of Aβ, it may not be taken up by this pathway (Bartnik et al., 2010).

The data demonstrate that short-term infusion with the D3 peptide leads to a reduction in the density of Aβ deposits in the hippocampus and cortex, whereas the infusion with the D1 peptide or the vehicle solution does not. Cognitive deficits are not significantly different between control and D3-infused mice. Furthermore, in the D3-infused mice, there is a significant reduction in both activated astrocytes and microglial cells surrounding Aβ plaques compared to the control and D1-treated mice. Together, this demonstrates that this peptide has great potential as a future treatment in AD.

REFERENCES

Akiyama, H., Barger, S., Barnum, S., Bradt, B., Bauer, J., Cole, G. M., et al. (2000). Inflammation and Alzheimer's disease. *Neurobiology of Aging, 21*(3), 383–421.

Bartnik, D., Funke, S. A., Andrei-Selmer, L. C., Bacher, M., Dodel, R., & Willbold, D. (2010). Differently selected D-enantiomeric peptides act on different Abeta species. *Rejuvenation Research, 13*, 202–205.

Bartolini, M., Bertucci, C., Bolognesi, M. L., Cavalli, A., Melchiorre, C., & Andrisano, V. (2007). Insight into the kinetic of amyloid beta (1-42) peptide self-aggregation: Elucidation of inhibitors' mechanism of action. *Chembiochem, 8*, 2152–2161.

Bondolfi, L., Calhoun, M., Ermini, F., Kuhn, H. G., Wiederhold, K. H., & Walker, L. (2002). Amyloid-associated neuron loss and gliogenesis in the neocortex of amyloid precursor protein transgenic mice. *Journal of Neuroscience, 22*, 515–522.

Braak, H., & Braak, E. (1991). Neuropathological stageing of Alzheimer-related changes. *Acta Neuropathologica, 82*, 239–259.

Braak, H., & Braak, E. (1998). Evolution of neuronal changes in the course of Alzheimer's disease. *Journal of Neural Transmission, 53*, 127–140.

Buxbaum, J. D., Thinakaran, G., Koliatsos, V., O'Callahan, J., Slunt, H. H., Price, D. L., et al. (1998). Alzheimer amyloid precursor protein in the rat hippocampus: Transport and processing through the perforant path. *Journal of Neuroscience, 18*, 9629–9637.

Dickson, T. C., & Vickers, J. C. (2001). The morphological phenotype of beta-amyloid plaques and associated neuritic changes in Alzheimer's disease. *Neuroscience, 105*, 99–107.

Drin, G., Cottin, S., Blanc, E., Rees, A. R., & Temsamani, J. (2003). Studies on the internalization mechanism of cationic cell-penetrating peptides. *Journal of Biological Chemistry, 278*, 31192–31201.

Franklin, K. B. J., & Paxinos, G. (1997). *The mouse brain in stereotaxic coordinates*. New York: Academic Press.

Funke, S. A., van Groen, T., Kadish, I., Bartnik, D., Nagel-Steger, L., Brener, O., et al. (2010). Oral treatment with the D-enantiomeric peptide D3 improves the pathology and behavior of Alzheimer's disease transgenic mice. *ACS Chemical Neuroscience, 1*, 639–648.

German, D. C., & Eisch, A. J. (2004). Mouse models of Alzheimer's disease: Insight into treatment. *Reviews in the Neurosciences, 15*, 353–369.

Haass, C., & De Strooper, B. (1999). The presenilins in Alzheimer's disease—Proteolysis holds the key. *Science, 286*, 916–919.

Hardy, J. (1992). An 'anatomical cascade hypothesis' for Alzheimer's disease. *Trends in Neurosciences, 15*(6), 200–201.

Hebert, L. E., Scherr, P. A., Bienias, J. L., Bennett, D. A., & Evans, D. A. (2003). Alzheimer disease in the US population: Prevalence estimates using the 2000 census. *Archives of Neurology, 60*, 1119–1122.

Hirtz, D., Thurman, D. J., Gwinn-Hardy, K., Mohamed, M., Chaudhuri, A. R., & Zalutsky, R. (2007). How common are the "common" neurologic disorders? *Neurology, 68*, 326–337.

Hyman, B. T., & Trojanowski, J. Q. (1997). Consensus recommendations for the postmortem diagnosis of Alzheimer disease from the National Institute on Aging and the Reagan Institute Working Group on diagnostic criteria for the neuropathological assessment of Alzheimer disease. *Journal of Neuropathology and Experimental Neurology, 56*, 1095–1097.

Ida, N., Hartmann, T., Pantel, J., Schröder, J., Zerfass, R., Förstl, H., et al. (1996). Analysis of heterogeneous A4 peptides in human cerebrospinal fluid and blood by a newly developed sensitive Western blot assay. *Journal of Biological Chemistry, 271*, 22908–22914.

Iijima, K., Liu, H. P., Chiang, A. S., Hearn, S. A., Konsolaki, M., & Zhong, Y. (2004). Dissecting the pathological effects of human Abeta40 and Abeta42 in Drosophila: A potential model for Alzheimer's disease. *Proceedings of the National Academy of Sciences of the United States of America*, *101*, 6623–6628.

Jankowsky, J. L., Slunt, H. H., Ratovitski, T., Jenkins, N. A., Copeland, N. G., & Borchelt, D. R. (2001). Co-expression of multiple transgenes in mouse CNS: A comparison of strategies. *Biomolecular Engineering*, *17*, 157–165.

Ji, Y., Permanne, B., Sigurdsson, E. M., Holtzman, D. M., & Wisniewski, T. (2001). Amyloid beta40/42 clearance across the blood-brain barrier following intra-ventricular injections in wild-type, apoE knock-out and human apoE3 or E4 expressing transgenic mice. *Journal of Alzheimer's Disease*, *3*, 23–30.

Kadish, I., & van Groen, T. (2002). Low levels of estrogen significantly diminish axonal sprouting after entorhinal cortex lesions in the mouse. *Journal of Neuroscience*, *22*(10), 4095–4102.

Kumar-Singh, S. (2008). Cerebral amyloid angiopathy: Pathogenetic mechanisms and link to dense amyloid plaques. *Genes, Brain, and Behavior*, 7(Suppl. 1), 67–82.

Lee, V. M., Goedert, M., & Trojanowski, J. Q. (2001). Neurodegenerative tauopathies. *Annual Review of Neuroscience*, *24*, 1121–1159.

Liu, H., Funke, S. A., & Willbold, D. (2010). Transport of Alzheimer disease amyloid-beta-binding D-amino acid peptides across an in vitro blood-brain barrier model. *Rejuvenation Research*, *13*, 210–213.

Liu, L., Ikonen, S., Heikkinen, T., Heikkilä, M., Puoliväli, J., van Groen, T., et al. (2002). Effects of fimbria-fornix lesion and amyloid pathology on spatial learning and memory in transgenic APP+PS1 mice. *Behavioural Brain Research*, *134*, 433–445.

Malm, T. M., Iivonen, H., Goldsteins, G., Keksa-Goldsteine, V., Ahtoniemi, T., Kanninen, K., et al. (2007). Pyrrolidine dithiocarbamate activates Akt and improves spatial learning in APP/PS1 mice without affecting beta-amyloid burden. *Journal of Neuroscience*, *27*, 3712–3721.

Masters, C. L., & Beyreuther, K. (2006). Alzheimer's centennial legacy: Prospects for rational therapeutic intervention targeting the Aβ amyloid pathway. *Brain*, *29*, 2823–2839.

Monaco, S., Zanusso, G., Mazzucco, S., & Rizzuto, N. (2006). Cerebral amyloidoses: Molecular pathways and therapeutic challenges. *Current Medicinal Chemistry*, *13*, 1903–1913.

Permanne, B., Adessi, C., Fraga, S., Frossard, M. J., Saborio, G. P., & Soto, C. (2002a). Are beta-sheet breaker peptides dissolving the therapeutic problem of Alzheimer's disease? *Journal of Neural Transmission. Supplementum*, *62*, 293–301.

Permanne, B., Adessi, C., Saborio, G. P., Fraga, S., Frossard, M. J., Van Dorpe, J., et al. (2002b). Reduction of amyloid load and cerebral damage in a transgenic mouse model of Alzheimer's disease by treatment with a beta-sheet breaker peptide. *The FASEB Journal*, *16*, 860–862.

Pluta, R., Misicka, A., Barcikowska, M., Spisacka, S., Lipkowski, A. W., & Januszewski, S. (2000). Possible reverse transport of beta-amyloid peptide across the blood–brain barrier. *Acta Neurochirurgica. Supplement*, *76*, 73–77.

Rogers, J., & Lue, L. F. (2001). Microglial chemotaxis, activation, and phagocytosis of amyloid beta-peptide as linked phenomena in Alzheimer's disease. *Neurochemistry International*, *39*, 333–340.

Selkoe, D. J. (2001). Alzheimer's disease: Genes, proteins, and therapy. *Physiological Reviews*, *81*, 741–766.

Selkoe, D. J., & Wolfe, M. S. (2007). Presenilin: Running with scissors in the membrane. *Cell*, *131*, 215–221.

Sheng, J. G., Price, D. L., & Koliatsos, V. E. (2002). Disruption of corticocortical connections ameliorates amyloid burden in terminal fields in a transgenic model of Abeta amyloidosis. *Journal of Neuroscience*, *22*, 9794–9799.

Soto, C., Kindy, M. S., Baumann, M., & Frangione, B. (1996). Inhibition of Alzheimer's amyloidosis by peptides that prevent beta-sheet conformation. *Biochemical and Biophysical Research Communications*, *226*, 672–680.

Stalder, M., Deller, T., Staufenbiel, M., & Jucker, M. (2001). 3D-Reconstruction of microglia and amyloid in APP23 transgenic mice: No evidence of intracellular amyloid. *Neurobiology of Aging*, *22*, 427–434.

Tanzi, R. E., & Bertram, L. (2005). Twenty years of the Alzheimer's disease amyloid hypothesis: A genetic perspective. *Cell*, *120*(4), 545–555.

Tanzi, R. E., Moir, R. D., & Wagner, S. L. (2004). Clearance of Alzheimer's Abeta peptide: The many roads to perdition. *Neuron*, *43*, 605–608.

Van Broeck, B., Van Broeckhoven, C., & Kumar-Singh, S. (2007). Current insights into molecular mechanisms of Alzheimer disease and their implications for therapeutic approaches. *Neurodegenerative Diseases*, *4*, 349–365.

van Groen, T., & Kadish, I. (2005). Transgenic AD model mice, effects of potential anti-AD treatments on inflammation and pathology. *Brain Research Reviews*, *48*, 370–378.

van Groen, T., Kadish, I., Wiesehan, K., Funke, S. A., & Willbold, D. (2009). In vitro and in vivo staining characteristics of small, fluorescent, Abeta42-binding D-enantiomeric peptides in transgenic AD mouse models. *ChemMedChem*, *4*, 276–282.

van Groen, T., Kiliaan, A. J., & Kadish, I. (2006). Deposition of mouse amyloid beta in human APP/PS1 double and single AD model transgenic mice. *Neurobiology of Disease*, *23*, 653–662.

van Groen, T., Liu, L., Ikonen, S., & Kadish, I. (2003). Diffuse amyloid deposition, but not plaque number, is reduced in amyloid precursor protein/presenilin 1 double-transgenic mice by pathway lesions. *Neuroscience*, *119*, 1185–1197.

van Groen, T., Miettinen, P., & Kadish, I. (2011). Transgenic AD model mice, effects of potential anti-AD treatments on inflammation, and pathology. *Journal of Alzheimer's Disease*, *24*, 301–313.

Wang, J., Tanila, H., Puolivali, J., Kadish, I., & van Groen, T. (2003). Gender differences in the amount and deposition of amyloidbeta in APPswe and PS1 double transgenic mice. *Neurobiology of Disease*, *14*, 318–327.

Wang, Y. J., Zhou, H. D., & Zhou, X. F. (2006). Clearance of amyloid-beta in Alzheimer's disease: Progress, problems and perspectives. *Drug Discovery Today*, *11*, 931–938.

Wiesehan, K., Buder, K., Linke, R. P., Patt, S., Stoldt, M., Unger, E., et al. (2003). Selection of D-amino-acid peptides that bind to Alzheimer's disease amyloid peptide abeta1-42 by mirror image phage display. *Chembiochem*, *4*, 748–753.

Wiesehan, K., Stöhr, J., Nagel-Steger, L., van Groen, T., Riesner, D., & Willbold, D. (2008). Inhibition of cytotoxicity and amyloid fibril formation by a D-amino acid peptide that specifically binds to Alzheimer's disease amyloid peptide. *Protein Engineering, Design & Selection*, *21*, 241–246.

Wiesehan, K., & Willbold, D. (2003). Mirror-image phage display: Aiming at the mirror. *Chembiochem*, *4*, 811–815.

Yoshiike, Y., Chui, D. H., Akagi, T., Tanakama, N., & Takashima, A. (2003). Specific compositions of amyloid-beta peptides as the determinant of toxic beta-aggregation. *Journal of Biological Chemistry*, *278*, 23648–23655.

AUTHOR INDEX

Note: Page numbers followed by "*t*" indicate tables.

B

C

D

E

F

G

H

K

L

M

N

O

Q

R

U

V

W

SUBJECT INDEX

Note: Page numbers followed by "*f*" indicate figures, and "*t*" indicate tables.

T

FSC
www.fsc.org
MIX
Papier | Fördert gute Waldnutzung
FSC® C083411

Kolibri 360 GmbH
Ferdinand-Jühlke-Straße 7
99095 Erfurt, Deutschland
produktsicherheit@kolibri360.de